THE AEF

TRAVELS

NO LONGER PROPERTY OF
ANYTHINK LIBRARIES/
RANGEVIEW LIBRARY DISTRICT

The

AERONAUTS

TRAVELS IN THE AIR

NO LONGER PROPERTY OF
ANYTHINK LIBRARIES/
RANGEVIEW LIBRARY DISTRICT

JAMES
GLAISHER

MELVILLE HOUSE
BROOKLYN • LONDON

The Aeronauts

Taken from *Travels in the Air* by James Glaisher,
Camille Flammarion, W. de Fonvielle and Gaston Tissandier,
originally published by Richard Bentley, London, 1871

Copyright © 2019 Melville House Publishing
Foreword copyright © 2019 by Professor Liz Bentley
All rights reserved.
First Melville House printing: September 2019

Melville House Publishing
46 John Street
Brooklyn, NY 11201

mhpbooks.com @melvillehouse

Design by Beste M. Doğan

A catalog record for this book is available from the Library of Congress
Library of Congress Control Number:2019946233

ISBN: 978-1-61219-796-8
ISBN: 978-1-61219-797-5 (eBook)

1 3 5 7 9 10 8 6 4 2

Printed in the United States of America

CONTENTS

CHAPTER VI

CHAPTER VII

CHAPTER VIII

CHAPTER IX

Foreword

James Glaisher could be described as a pioneer of scientific meteorology. His drive to improve the quality of atmospheric measurements and the establishment of a network of reliable meteorological observers across the British Isles during the mid-1800s led to a climatological data set that was used for scientific research—although Glaisher himself was probably better known for risking life and limb during pioneering work to measure the atmosphere at altitude, squeezed into a wicker basket dangling under a hot air balloon and surrounded by an array of instruments.

James Glaisher was a founding member the British (now Royal) Meteorological Society in 1850. He was the society's mainstay in its early years, acting as secretary from its inception until 1872, president between 1867 and 1868 and editor of its many publications. He even named his elder son James Whitbread Lee Glaisher after the first and third presidents of the society, Samuel Whitbread and John Lee.

His enthusiasm and drive at the Royal Meteorological Society led him into roles at other institutions, including the Royal Society, Royal Astronomical Society, Royal Photographic Society, Royal Microscopical Society and the Royal Aeronautical Society.

His interest in scientific instruments started at an early age in his father's workshop and through having access to them at the Royal Observatory at Greenwich. This led to a desire for ensuring accurate observations that remained with Glaisher throughout his life.

By his early twenties he was working for the Ordnance Survey, in charge of meteorological observations in Ireland. This may have led to his interest in the atmosphere at higher levels since most of his work was on the summits of the Bencorr and Keeper mountains.

In 1840, a Magnetical and Meteorological Observatory became fully operational at Greenwich and Glaisher was appointed supervisor. At this time scientific meteorology was in its infancy. Glaisher was determined to make a name for himself in this new field of research and put all his energy into the duties of the new post.

Glaisher established a network of amateur observers, visiting them all personally to verify all their instruments before they were issued and measure any errors in them. This network was the beginning of a widespread system of simultaneous observations across the British Isles and would lead to an international network in the years to come. In 1849 he started a series of quarterly reports on the meteorology of England, published by the Registrar General for more than half a century.

In 1862 the British Association for the Advancement of Science (now the British Science Association) decided to fund a series of flights to study the upper atmosphere. The balloons would fly as high as possible, to heights of 37,000 feet or seven miles above the ground. Glaisher volunteered to perform these potentially dangerous flights. In all he made twenty-eight ascents between 1862 and 1866, thirteen of which were funded by the Association. His usual pilot was the experienced balloonist Henry Coxwell.

✦

Glaisher wanted to measure the atmosphere at high altitudes in order to gain insights into the temperature and humidity at different elevations. The scientific results were published in five reports submitted to annual meetings of the British Association for the Advancement of Science and Glaisher also regularly reported accounts of his ascents to the public through newspaper articles.

You may be familiar with James Glaisher's story from the film *The Aeronauts*—although the film replaces Henry Coxwell with a fictitious female character, Amelia Wren. This edition of *The Aeronauts*, taken from the book *Travels in the Air* originally published in 1871, tells of Glaisher's story and research in his own words. It demonstrates his determination, scientific prowess and fascination with technology and scientific instruments, which helped lay the groundwork for today's understanding of meteorology. His writing also beautifully captures his aerial journeys, describing the world from a bird's-eye view that so few people in the nineteenth century got to experience for themselves. As he writes:

"We seem to be citizens of the sky, separated from the earth by a barrier which seems impassable [. . .] In the upper world [. . .] the silence and quiet are so intense that peace and calm seem to reign alone."

—Professor Liz Bentley, Chief Executive at the Royal Meteorological Society

Temperature Conversion Chart

All temperatures depicted in the main body of text throughout the book are in Fahrenheit.

Celsius (°C)	Fahrenheit (°F)	Celsius (°C)	Fahrenheit (°F)
-50 °C	-58.0 °F	0 °C	32.0 °F
-40 °C	-40.0 °F	1 °C	33.8 °F
-30 °C	-22.0 °F	2 °C	35.6 °F
-20 °C	-4.0 °F	3 °C	37.4 °F
-10 °C	14.0 °F	4 °C	39.2 °F
-9 °C	15.8 °F	5 °C	41.0 °F
-8 °C	17.6 °F	6 °C	42.8 °F
-7 °C	19.4 °F	7 °C	44.6 °F
-6 °C	21.2 °F	8 °C	46.4 °F
-5 °C	23.0 °F	9 °C	48.2 °F
-4 °C	24.8 °F	10 °C	50.0 °F
-3 °C	26.6 °F	20 °C	68.0 °F
-2 °C	28.4 °F	30 °C	86.0 °F
-1 °C	30.2 °F	40 °C	104.0 °F

INTRODUCTION

I have elsewhere expressed my opinion that the balloon should be received only as the first principle of some aerial instrument which remains to be suggested. In its present form it is useless for commercial enterprise, and so little adapts itself to our necessities that it might drop into oblivion tomorrow, and we should miss nothing from the conveniences of life. But we can afford to wait, for already it has done for us that which no other power ever accomplished; it has gratified the desire natural to us all to view the earth in a new aspect, and to sustain ourselves in an element hitherto the exclusive domain of birds and insects. We have been enabled to ascend among the phenomena of the heavens, and to exchange conjecture for instrumental facts, recorded at elevations exceeding the highest mountains of the earth.

Doubtless among the earliest aeronauts a disposition arose to estimate unduly the departure gained from our natural endowments, and to forget that the new faculty we had assumed, while opening the boundless regions of the atmosphere as fresh territory to explore, was subject to limitations a century of progress might do little to extend. In the time of [Vicenzo] Lunardi, a lady writing to a friend about a balloon voyage she had recently made, expresses the common feeling of that day when she says that "the idea that I was daring enough to push myself, as I may say, before my time, into the presence of the Deity, inclines me to a species of terror"—an exaggerated sentiment, prompted by the admitted hazard of the enterprise (for [Jean-François] Pilâtre de Rozier had lately perished in France, precipitated to the earth by the bursting of his balloon), or dictated by an exultant and almost presumptuous sense of exaltation: for the first voyagers in the air, reminded by no visible boundary that for a few miles only above the earth can we respire, appear to have forgotten that the height to which we can ascend and live has so definite a limitation.

But no method more simple could have been imagined than that by which the aeronaut ascends, and which leaves the observer entire freedom to note the phenomena by which he is surrounded. With the ease of an ascending vapour he rises into the atmosphere, carried by the imprisoned gas, which responds with the alacrity of a sentient being to every external circumstance, and lends obedience to the slightest variation of pressure, temperature, or humidity. The balloon when full and on the earth, with a strong wind, is vehemently agitated, and if a stiff breeze prevails during the progress of inflation, it is for the time almost ungovernable. When prepared for flight it offers the greatest powers of resistance to mechanical control, and, bent on soaring upwards, struggles impatiently to be free.

In a line of perpendicular ascent the balloon has a motion of its own. It therefore rises or falls according to the action of the atmosphere upon the imprisoned gas. The second motion, which, united to the first, carries the balloon out of the perpendicular line on rising, and directs its onward motion in a plane, is not inherent in the balloon, but is due to the external force of horizontal currents which sweep it in the direction of their course, and communicate a compound motion we can neither direct nor calculate. The simple inherent motion we can repeat at will.

I believe the most timorous lose their sense of fear as the balloon ascends and the receding earth is replaced by the vapours of the air; and I refer this confidence chiefly, as has been suggested, to the consciousness of isolation by which the balloon traveller feels more like a part of the machine above than of the world below. Thus situated, he is induced to forget the imperfections of the machine in witnessing the close accordance of its movements with those of the surrounding clouds. The balloon strives to attain a height where it may rest in equilibrium with the air in which it floats; its ascent is checked by allowing gas to escape by the valve, and by the weight of ballast, but facilitated by keeping the gas in and discharging the ballast. These are the methods by which it is made to rise or fall at the will of the aeronaut, and the only objection to the frequent employment of the valve and the use of ballast is to be found in the greatly abbreviated life of the balloon and too rapid diminution of its powers which follow.

Up to the time of the balloon we had no means of ascending by which we could test the conditions of the atmosphere for even a mile above the surface of the earth, apart from the terrestrial influences and the inevitable labour of ascending the mountain-side. When, therefore, Messrs [Jacques] Charles and

[Nicolas-Louis] Robert made their first ascent, and recorded the history of their sensations and the conditions of the atmosphere at various elevations, as the natural incidents and circumstances of their voyage, a practical application of the balloon was thus spontaneously suggested.

Before [Joseph Louis] Gay-Lussac solicited the French government for the use of the balloon in which he ascended to the height of 23,000 feet, M. [Horace Bénédict] de Saussure, of Geneva, had alone made observations at a height of 15,000 feet and upwards; a distinction he had won by accomplishing the desire of his life, and ascending to the summit of Mont Blanc.

This memorable journey de Saussure performed in the summer of 1787, four years after the first balloon ascent of Messrs Robert and Charles in a hydrogen balloon from Paris, and seventeen years before Gay-Lussac made his ascent for the advancement of science. The weather was favourable, and the snow compact and hard. Accompanied by his servant and eighteen guides, de Saussure began his journey. There was no difficulty or danger in the early part of the ascent, their footsteps being either on the grass or the rock itself. After six hours' incessant climbing, they found themselves 6,000 feet above the village of Chamouni [Chamonix], from which they started, and 9,500 feet above the level of the sea. At this height, the same to which M. Robert had attained in his balloon, de Saussure and his party prepared to encamp, and slept under a tent on the edge of the glacier of the Montagne de la Côte. By noon the next day they were 2,000 feet above the level of perpetual frost. In the afternoon, after eight hours of climbing, they had arrived at an elevation of 13,300 feet above the level of the sea. They were now on the second of the three tremendous steps which extend from 800 to 1,000 feet each between Les Grands Mulets and the summit of Mont Blanc. On the second of Les Mulets, de Saussure intended to pass the night. The guides dug out the snow for their lodging, and threw some straw into the bottom of the pit, across which they stretched a tent. Their water was frozen, and they had but a small charcoal brazier, which proved quite insufficient to melt snow for twenty persons. When morning came, they prepared again for departure. The cold was excessive, but before breakfast could be obtained it was necessary to melt the show which also served for the water in their journey to come. They crossed the great ice plain, or Grand Plateau, without difficulty; but the rarefaction of the air began to affect their lungs, and this inconvenience continued to increase at every step. A prolonged rest was made in hopes of recruiting their forces, but with little advantage. They had not gone a dozen steps before they were compelled to halt to recover breath, and in this manner, slowly and with great toil and discomfort, the summit was reached.

"At last," writes de Saussure, "I had arrived at the long-wished-for end of my desires. As the principal points in the view had been before my eyes for the last two hours of this distressing climb, almost as they would appear from the summit, my arrival was by no means a *coup de théâtre*; it did not even give me the pleasure that one might imagine. My keenest impression was one of joy at the cessation of all my troubles and anxieties: for the prolonged struggle and the recollection of the sufferings this victory had cost me produced rather a feeling of irritation. At the very instant that I stood upon the most elevated point of the summit, I stamped my foot on it more with a sensation of anger than pleasure. Besides, my object was not only to reach the crown of the mountain: I had to make such observations and experiments as alone would give any value to the enterprise, and I was afraid I should only be able to accomplish a portion of my intentions. I had already found out, even on the plateau where we slept, that every careful observation in such a rarefied atmosphere is fatiguing, because the breath is held unconsciously; and as the tenuity of the air is obliged to be compensated for by the frequency of respiration, this suspended breathing causes a sensible feeling of uneasiness. I was compelled to rest and pant as much, after regarding one of my instruments attentively, as after having mounted one of the steepest slopes."

De Saussure spent three hours and a half in observations, and after four hours passed on the summit, began with his party to descend. They passed the night on Les Mulets, the third since they left Chamouni, and de Saussure writes: "We supped merrily together and with famous appetites. It was not until then that I really felt pleased at having accomplished the wish of twenty-seven years. At the moment of my reaching the summit I did not feel really satisfied. I was less so when I left it: I only reflected then upon what I had *not* done. But in the stillness of the night, after having recovered from my fatigue, when I went over the observations I had made; when especially I retraced the magnificent expanse of the mountain peaks, which I had carried away engraven on my mind; and when I thought I might accomplish on the Col de Géant what most assuredly I should never do on Mont Blanc, I enjoyed a true and unalloyed satisfaction." The simple narrative of this eminent man is throughout a commentary upon the use of the balloon for the purpose of vertical ascent. To be carried up with speed and certainty at any number of feet per minute, with instruments complete and carefully prepared for observation, the observer seated as calmly as in his observing room at home, are advantages which speak for themselves. The observations of today can be repeated tomorrow, and successively throughout the seasons of the year, and at different hours of the day; and the importance of this repetition is rendered clear by considering of what slight value is a single set of observations, whether in meteorology or any other branch of inquiry, except to appease curiosity, and how little gain to science is one isolated day's experience; and yet to ascend Mont Blanc was the one great fact of de Saussure's life.

The view which offers itself to an aeronaut seated conveniently in the car of a balloon, is far more extended than any the eye can embrace within its scope from the summit of a lofty mountain. It is gained without fatigue, but then there is no succession of magnificent scenery which compensates for the toil of the Alpine traveller, and suggests a variety of observations unknown to the voyager of the atmosphere. To the latter, situated at a height above the earth, separated from all communication with it, the scenery on its surface is dwarfed to a level plane, and the whole country appears like a prodigious map spread out beneath his feet. Better than the Alpine traveller he can trace the history of physiological sensations, and pursue the observations of meteorology. In the one case he travels free from the effects of muscular exertion, which makes fatigue so formidable in the higher regions of the earth's scenery, and, apart from all terrestrial influences of soil and temperature, scan the true conditions of the atmosphere.

On looking into the annals of aerostation, I do not find that balloon travellers in general have cared to ascend beyond the height to which de Saussure attained on the summit of Mont

Blanc, and the greater number of ascents are within this limit. Most aeronauts have taken care to keep well within recognition of the visible scenery of the earth, and would seem to have been too eager to enjoy the privilege of movement, and the varied prospect in any direction they could travel, to wish to prove their capacity for vertical ascents. We have few reliable observations to a great height. High ascents have now and then been attempted by professional aeronauts eager to gain the attention of the public and enlist its sympathy in their results. Voyages in illuminated balloons by night, in weather not always suitable, were performed successively by M. [Jean-Pierre] Blanchard, and after him by M. [André-Jacques] Garnerin, who preceded the late Mr. [Charles] Green. Beyond the passing sensation of the moment, recorded in the public prints of the day, their ascents have left no permanent trace in the history of the balloon. The ascent made by M. Charles, after a joint expedition of Messrs Charles and Robert, is the first experience of value we have to compare with others. It was, we may suppose, the first occasion on which sunset was witnessed a second time in the same day by any living mortal.

On December 1, 1783, having descended and landed his companion, M. Charles determined to ascend alone. It was towards sunset, and ballast could not be readily procured. Without waiting, therefore, M. Charles gave the signal to the peasants, who were holding his machine, to let go; "and I sprang," says M. Charles, "like a bird into the air. In twenty minutes I was 1,500 toises* high, out of sight of terrestrial objects. The globe,

* A unit of measurement originating in pre-revolutionary France; 1,500 toises is roughly 2,900 metres.

which had been flaccid, swelled insensibly; I drew the valve from time to time, but still continued to ascend. For myself, though exposed to the open air, I passed in ten minutes from the warmth of spring to the cold of winter: a sharp, dry cold, but not too much to be borne. In the first moment I felt nothing disagreeable in the change. In a few minutes my fingers were benumbed by the cold, so that I could not hold my pen. I was now stationary as to rising and falling, and moved only in a horizontal direction. I rose up in the middle of the car to contemplate the scenery around me. When I left the earth, the sun had set on the valleys; he now rose for me alone; he presently disappeared, and I had the pleasure of seeing him set twice on the same day. I beheld for a few seconds the circumambient air, and the vapours rising from the valleys and rivers. The clouds seemed to rise from the earth, and collect one upon the other, still preserving their usual form, only their colour was grey and monotonous from the want of light in the atmosphere. The moon alone enlightened them, and showed me that I had changed my direction twice. Presently I conceived, perhaps a little hastily, the idea of being able to steer my course. In the midst of my delight I felt a violent pain in my right ear and jaw, which I ascribed to the dilatation of the air in the cellular construction of those organs as much as to the cold of the external air. I was in a waistcoat and bare-headed; I immediately put on a woollen cap, yet the pain did not go off till I gradually descended."

M. de [Jean Baptiste Marie Charles] Meusnier [de la Place] made various calculations as to the height attained by M. Charles, and calculated it to have been at least 9,000 feet. The temperature at the time of starting was 47° on the earth, but in ten minutes had descended to 21°. When M. Charles came down and landed his companion, they were met by the Duc de Chartres and some French noblemen, who had followed on horseback for twenty miles the course of the balloon. A contemporary pamphlet records the particulars of the ascents, and has a postscript to the effect that Messrs Charles and Robert were arrested on returning to Paris, by order of the King, who, at the suggestion of two of his ecclesiastics, adopted this course to prevent the further endangering the lives of his subjects. "But," adds the writer of the pamphlet, "as great interest is making for them, it is thought they will speedily be discharged."

The height to which M. Charles ascended was thought to be enormous. There had been nothing like it before, and this, the first essay of the hydrogen balloon, brought it at once into public favour and notice. The same elevation attained one year and ten months later upon the mountain side, made de Saussure console himself under failure, with the thought that he had made more valuable barometric observations and had been higher than any other traveller in Europe. On this occasion he had attempted to ascend Mont Blanc; but the route to the summit remained undiscovered, and after journeying for a day his party were forced to return. Passing the night at an elevation of 9,000 feet, within the walls of a rude hut which had been constructed for the expedition, de Saussure gained his first impressions of these elevated regions. Two mattresses had been deposited within the hut, and an open parasol set against the entrance formed the door. De Saussure says: "As night came on, the sky was completely pure and cloudless; the stars, brilliant indeed, but unscintillating, cast a pale light over the summit of the mountain peaks, sufficient to define their size and distance. The repose and dead silence which reigned in this immeasurable space, increased by the imagination, inspired me almost with terror. It appeared as though I was left living alone in the world, and that I saw its corpse at my feet. I either slept lightly and calmly, or my thoughts were so bright and peaceful I was sorry to slumber. When the parasol was not before the door I could see from my bed the snow, the ice, and the rocks below the cabin, and the rising of the moon gave the most singular appearance to the view." Some of the party who shared the hut with de Saussure suffered greatly from the rarefaction of the air, and could not eat anything. The next morning, after an hour's climb, they were forced to return. The snow was soft, and they encountered treacherous drifts and blocks of ice. De Saussure therefore with reluctance abandoned his attempt, the last which was made before the discovery of the true route to the summit.

Whether by mountain ascents or balloon voyages, the traveller who quits the ordinary level of the earth for the upper regions finds two inevitable conditions presented to his endurance, arising respectively from the gradual loss of heat, and the tenuity of the atmosphere. The effects of these conditions will differ, we may assume, with every individual, but certainly are more uniform in their relation to the occupant of a balloon car, who is spared the necessity of exertion and consequent fatigue, than the effects of similar conditions upon a mountain traveller, who, to attain a height to which the aeronaut can ascend in an hour, is subjected to the continuous toil of two successive days, devoted to an ascent which is granted only to a certain degree of strength and activity; for of those who have attempted to reach the summit of Mont Blanc, many have failed through physical inability to endure fatigue. The test, therefore, of the rigorous severity of the upper regions has been experienced by those chiefly of more than average *physique*, men equal to the toil, and who have kept themselves in previous training for the severe exercise involved in the undertaking. But the aeronaut enters upon his expedition unprepared, and attains an elevation not dependent on his physical strength. To this cause, probably, balloon voyages under apparently similar circumstances of elevation, show results by no means uniform: a fact which has provoked severe criticism, and has been supposed to arise from the vanity of individuals wishing to prove their experiences greater than those of others. I should be sorry to have to be the champion of all the marvellous histories that have been related; but on looking over a collection of narratives from 1783 to 1835, including the principal aerostatic voyages performed in England, Italy, and France, I believe that, as a rule, authors have written their true experiences, and have correctly recorded their impressions. Aeronauts by trade may at times have been guilty of exaggeration, but the tyro who ascends once and never again is most likely to make demands upon our credulity. It happens thus—that the diminished pressure of the air, and the unfamiliar circumstances of his position, act with far greater force upon an individual who ascends for the first time than ever afterwards. This I can attest, having ascended without the slightest inconvenience to a height which used to produce discomfort, and even discoloration of the hands and face, until at length I became so acclimatized to the effects of a more rarefied atmosphere, that I could breathe at an elevation of four miles at least above the earth without inconvenience, and I have no doubt that this faculty of acclimatization might be so developed as to have a very important bearing upon the philosophical uses of balloon ascents. At six and seven miles high, I experienced the limit of our power of breathing in the attenuated atmosphere. More frequent experiments would increase this height, I have little doubt, and artificial appliances might be contrived to continue it higher still. A boundary must exist, but I have little hesitation in saying that it might be removed beyond its present limit. To the terrestrial traveller the conditions of diminished heat and increased tenuity of atmosphere present themselves in the light of problems which have more relation to the influences of the earth than of the atmosphere. Clinging to the earth at every step, and completing his journey upon the highest point of his terrestrial pinnacle, he cannot clear his observations from the influences of the earth; or mark the gradual diminution of temperature conjointly with the amount and degree of cloud present, and estimate, by repeated observations, the extent to which the latter serves as a radiating screen to keep back the heat of the earth within the limits of the lower atmosphere. He cannot mark the fluctuations of temperature through which he rises on a fine but cloudy day, and make them comparable with others taken during cloudless ascents, with no local disturbing causes present to interfere with the law of a decreasing temperature with increase of height. These belong to the balloon voyager alone.

As a rule, the toil of a terrestrial ascent has induced the painful sensations of a rarefied air at an elevation where the aeronaut would have sat at ease, with little or but trifling inconvenience. Thus, at the height to which M. Charles ascended and felt but a slight pain in the muscles of his face and discomfort in his ears, M. Bouret, the friend of de Saussure, suffered so keenly that he was compelled to descend. At a height of three miles I never experienced any annoyance or discomfort; yet there is no ascent, I think, of Mont Blanc in which great inconvenience and severe *pain* have not been felt at a height of 13,000 feet; but then, as before remarked, this is an elevation attained only after two successive days of toil. About this elevation, Dr [Joseph] Hamel and his party, having passed the Grand Plateau, speak of incessant thick and laboured respiration. They returned, however, without reaching the summit, appalled by the catastrophe of an avalanche of snow, which hurried three of the guides into the frightful depths of a crevasse on the ascending slope of Le Mont Maudit. This fatal attempt was made in the early part of the present century, Dr Hamel being anxious to make the ascent in furtherance of some especial observations taken in compliance with instructions received from the Emperor of Russia. Later still, the same ground was passed over by a party including Sir Francis Talfourd and his son. The effects of cold and diminished pressure are clearly shown in the narrative which is elsewhere published. "The line of our march," observes Sir Francis, "lay up long slopes of snow ascending in a steep inclination before us. There was nothing to vary the toil or the pain, except that as fatigue crept on, and nature began to discriminate between the stronger and the weaker, our line was no longer continuous, but broken into parties. The rarity of the atmosphere now began to affect us, and, as the disorder arising from this cause was more impartial than the distribution of muscular activity, our condition was for a time almost equalized. Violent nausea and headache were experienced by one of our party, while I only felt, in addition to the distress of increasing weakness, the taste or scent of blood in the mouth, as if it were about to burst from the nostrils. We thus reached the Grand Plateau, a long field of snow in the bosom of the highest pinnacles of the mountain."

Until the aeronaut shall have found means to ascend beyond the present limit, he will, I believe, feel no sensation of cold so painful as that of the Alpine traveller. At the extreme height to which I have ascended, the lowest temperature was 12° below zero, or 44° below the freezing point of water. The cold was intense, but not painfully severe, and no amount of suffering was experienced from this cause; of five pigeons taken up, but one perished. All authorities agree that cold, however intense, is supportable under a calm temperature, whereas a moderate degree of cold with a fresh breeze, or the slightest air stirring, produces the sensation of a very low temperature. The balloon voyager, who feels no wind because he always travels with it, and when sweeping along with the speed of an express train yet meets no current as he cleaves the air and knows no motion, can bear the cold to which he is subjected with little demand on his power of endurance. It is true he is condemned to immovability and to vicissitudes of cold both dry and wet, but these extremes can be guarded against by due precautions of fur and warm clothing. During the period of his voyage the aeronaut may create defence enough against the fluctuations of the atmosphere.

The subject of cold, physiologically considered with regard to our own sensations, M. [Charles] Martins has ably treated in his essay, "Du froid thermométrique et de ses relations avec le froid physiologique," in plains and mountains. "Of those who suffer death from cold," M. Martins writes, "let us suppose a single traveller, or a small caravan, wishing to cross one of the *'Cols'* covered with eternal snow which lead from Valais to Piedmont, or from France to Spain. It is winter, or the commencement of spring, or the end of autumn. The journey is long, the time uncertain. The voyagers are not perfectly acquainted with the country. They set out. The sky is covered with cloud, which descends little by little, and envelops them in a thick mist. They walk in the snow, in the track of those travellers who have preceded them; but soon other traces cross those by which they guide themselves, or a recent fall of snow has obliterated every mark. They stop, hesitate, return upon their steps, turn themselves sometimes to the right, sometimes to the left, always making for a summit; they can scarcely see through the fog and mist. The snow begins to fall, not flaky as on the plains, but granulated, dry, and like hail. Driven by the wind, it penetrates to the skin through the strongest vestments; striking incessantly the face, it produces a permanent giddiness which soon becomes vertigo. Then the poor traveller, worried, harassed, and not seeing two steps before him, feels an irresistible desire to sleep. He knows that sleep is death; but, lost and despairing, he seeks some rock, and abandoning himself lies down to rise no more. His pulse declines as in a lethargy, and he dies of cold, as one dies of inanition. Moral energy in these moments is the only means of safety. It is necessary at all risks to combat sleep, to walk, to defend oneself against the cold by muscular exercise."

"Jacques Balmat, who was the first to make the ascent of Mont Blanc," observes M. Martins, "knew it well. He was left alone on the Grand Plateau. There he was surprised by night: to mount to the summit was impossible; to re-descend in the obscurity equally impossible. He took his post valiantly, and walked about the snow till morning." This man was a native of Chamouni, and had accompanied the party of Dr. [Michel-Gabriel] Paccard. Being, it is supposed, at the time unpopular among his comrades, he had been neglected by them during the ascent: when they decided to return, he had lost sight of them, and his companions, either forgetful of him or determined to descend without him, had returned upon their steps, and he found himself, at an elevation of 14,000 feet, abandoned in the midst of a blinding storm of snow, without food, and but poorly clad. Half dead from the piercing cold, his limbs numbed by the labours he had undergone, the poor fellow passed this terrible night as best he could. When morning dawned, Balmat decided upon his part; his feet were frostbitten and had lost all sensation; but his limbs, benumbed and paralysed, he resolved should carry him to the summit never before attained. Alone he accomplished that which had been denied his treacherous comrades. Alone he traversed the untrodden fields of snow, climbed hitherto inaccessible slopes of ice, and forced his way to the summit by a route but little changed up to the present time. That evening he returned to his village, and, prostrate and despairing of his life, submitted himself to the services of Dr Paccard, the physician of Chamouni. After an illness of several weeks, in gratitude to the doctor he revealed to him in confidence his secret; and when Balmat was sufficiently recovered, he and Paccard made the first ascent together. They were delighted with their success, and wrote at once to de Saussure at Geneva, who immediately ordered an equipment of mules and guides, to be accompanied and attended by porters and attendants. With the first favourable opportunity of the season, de Saussure made his celebrated ascent, as we have related, Jacques Balmat being appointed chief of the troop of guides. This is the popular narrative, and, to his moral energy alone Balmat owed his preservation from death on the night that he was exposed to the piercing and insidious cold of so great an elevation.

A very rapid descent is productive of inevitable discomfort. To this cause probably M. Robert owed the severe pain and inconvenience he experienced at 9,000 feet. The year following, Messrs Charles and Robert ascended to a height of 14,000 feet. In March 1784, M. Blanchard, the celebrated French aeronaut, made his first ascent from Paris; he mounted high above the clouds, and attained the elevation of 9,600 feet. There is no mention in either case of personal inconvenience. Messrs [Louis-Bernard Guyton de] Morveau and [Claude-Philippe] Bertrand ascended from Dijon in April 1784, when they attained the height of 13,000 feet, and travelled eighteen miles in twenty-five minutes. The temperature of the air descended to 25°. In June 1784, M. Fleurant and Madame [Elisabeth] Thible ascended at Lyons in a very large fire balloon, named *Le Gustave*, before the King of Sweden. They reached the height of 8,500 feet, and travelled only two miles in forty-five minutes. In Signor Lunardi's balloon, Mrs [Letitia Ann] Sage ascended with Mr. [George] Biggin from London; in kneeling down to secure the fastenings of the network in the opening of the gallery, the lady broke the barometer, and they had no measure therefore of the height to which they ascended. It was, however, considerable.

In July 1784, M. Robert ascended from Paris with the Duc de Chartres and other gentlemen. Within the hydrogen balloon was enclosed a smaller one, filled with common air. They ascended to a height of 5,100 feet, and were greatly beaten about by an eddy or revolving current. The gas expanded; they had no valve, and the inner balloon choked up the aperture of the neck and permitted no escape. In this dilemma, at the mercy of a whirlwind, they decided to make a rent in the outer covering. The Duc de Chartres himself took one of the banners and made two holes in the balloon, which formed an aperture between seven and eight feet in length. The gas escaped in volumes through the open rents, and they came down with great velocity, but no one was injured.

In September 1784, Signor Vincenzo Lunardi ascended, taking with him one small thermometer. He attained no considerable elevation. In January 1785, M. Blanchard and Dr [John] Jeffries crossed the Channel in a hydrogen balloon from Dover to Calais. From some defect in the gas, or deficiency in its amount, far from being affected by the rarity of the air, they could with difficulty keep themselves at a level above the sea, and to do so were obliged to part with everything in the car, and even take off their clothes and throw them overboard. As they neared the land, however, the balloon rose, and, describing a magnificent arch, carried them over the high ground surrounding Calais, and finally landed them in the Forest of Guiennes. On July 22, Signor Lunardi ascended from Liverpool. The process of filling the balloon was tedious, and the impatience of the populace made it necessary to ascend before the process of inflation could be properly performed. He therefore found himself with barely enough rising power to carry him, and without ballast of any kind; so that when after being becalmed he was gently wafted towards the sea, he had not ballast to throw out to enable him to rise and meet some other current. When suspended over the sea, to lighten his weight he threw down his hat, upon which the balloon rose, and the thermometer fell 3°. The balloon entered a cloud, and, with the thermometer at 50°, Lunardi was surprised at finding himself surrounded with a shower of snow. Being desirous to ascend higher, he threw down his banner, and shortly after took off his coat (the uniform of the Honourable Artillery Company) and threw it away. He then rose majestically, and bore towards the land. Ten minutes later he perceived a thunder cloud, and signs of a gathering storm. To pass from its vicinity he threw down his waistcoat. The temperature had fallen to 32°, and five minutes later fell to 27°; the snow had melted on the top of his balloon, and had trickled down in the form of water. It was now congealed in the colder temperature, and hung in icicles round the neck of the balloon; he shook off about a pound's weight, and it fell upon the floor of the gallery, Lunardi looking upon it as the ballast of Providence. The temperature descended to 26°. He now began to descend. It was three minutes to seven, and six minutes after he was safely landed in a cornfield about twelve miles from Liverpool. Here we have a practical commentary upon the necessity of a proper freight of ballast, and of a nicely regulated equilibrium between the balloon and surrounding atmosphere before starting. In the month preceding, M. Pilâtre de Rozier and M. [Pierre] Romain had made their last and fatal voyage from Boulogne. The balloon employed was compound, a small fire balloon being appended to a hydrogen balloon above. The one set fire to the other, and the aeronauts were precipitated to the earth and killed.

In the beginning of the next century the name of M. Garnerin is closely associated with balloon history, and replaces that of M. Blanchard. He is chiefly memorable for night ascents with an illuminated balloon. On July 5, 1802, M. Garnerin ascended from Marylebone; the wind was high, but he rose to a height of 7,800 feet, and descended at Chingford, near Epping Forest. His fame as an aeronaut was considerable, and his popularity about this time was at its culminating point with the people of the metropolis, who were in a state of tumult to witness his ascent. This was his twenty-seventh voyage in Europe.

In 1804, Professor [Etienne-Gaspard] Robertson ascended from St. Petersburg, accompanied by the academician, Sacharof. This was purely a scientific voyage, instituted at the request of the Russian Academy, to ascertain the physical state of the atmosphere, and the component parts of it at different determinate heights; also the difference between the results given by vertical ascent and the observations of [Jean-André] de Luc, de Saussure, [Alexander von] Humboldt, and others, on mountains, which it was rightly concluded could not be so free from terrestrial influences as those made in the open air. Among the experiments proposed by the Academy which were to be made at great distances from the earth, the following were included: The change of rate of evaporation of fluids; the decrease or increase of the magnetic force; the inclination of the magnetic needle; the increase of the power in the solar rays to excite heat; the greater faintness of the colours produced by the prism; the existence or non-existence of electric matter; observations on the influence and changes which the rarefaction of the air occasions in the human body; the flying of birds; the filling with air of exhausted flasks, at each fall of an inch, in the barometer; and some other chemical and philosophical experiments.

These are the questions to which every voyager in behalf of science is required to add some testimony in reply. In the case of Mr. Robertson, the gyrating movement of the balloon was a difficulty, as it is to all aeronauts, and rendered observations with the deflecting needle almost impossible. With the barometer at 27 inches, Mr. Robertson and M. Sacharof experienced no more inconvenience than a numbness of sensation in their ears, and no alteration of sound, which at 23 inches was the same as on the earth's surface. At the height of 22 inches they were nearly surrounded by fog, the earth appearing enveloped in a smoke-coloured atmosphere which a good telescope failed to penetrate.

Having discharged their ballast and thrown down every available article from the car of the balloon, they deposited for safety their instruments in the centre of a bundle made of their warm clothing, and lowered it together with their grapnel. This proceeding was intended to obviate the breakage consequent on a rough descent. The balloon, so lightened on descending, flew up again to the limit of the cord, but soon effected a safe and gentle landing. The instruments, roughly dragged along the surface of the ground with the package of which they formed a part, were, as might have been expected, injured or broken. These gentlemen made various minute observations of interest, and intelligently recorded all that they witnessed during their ascent. But the instruments could not easily be used in the car of a balloon, and the results required confirmation by subsequent experiments; opportunities also were lost by fog and a clouded atmosphere, and the practical embarrassments of balloon management were severely felt; so that the results are meagre, and show the necessity of system and repeated practice to arrive at results of value.

On October 7, 1803, Count [Francesco] Zambeccari, Dr [Gaetano] Grassati, of Rome, and M. Pascal Andreoli, of Ancona, made a night ascent in a fire balloon from Bologna. They took with them instruments, and a lantern by which to see to make observations. The balloon rose with great velocity, and soon attained a height at which Count Zambeccari and Dr Grassati became insensible. M. Andreoli retained the use of his faculties.

About two in the morning they found themselves descending over the waves of the Adriatic; the lantern had gone out, and to light it was a work of no little difficulty. The balloon continued to descend rapidly, and fell, as they anticipated, into the sea. Thoroughly drenched, they succeeded in throwing out ballast until they rose again, and passed through three successive regions of cloud, which covered their clothes with rime, and in this situation they became deaf, and could not hear each other speak. About three o'clock the balloon again descended, and was driven by a gust of wind to the coast of Istria, bounding in and out of the sea till eight o'clock in the morning, when one Antonio Bazon picked them up in his ship, and carried them to shore. The balloon, left to itself, went over to the Turks, having first mounted to an amazing height. The most intense interest was excited for the fate of the aeronauts, and bulletins of health were sent from Venice to Bologna. Count Zambeccari suffered most, and was forced to have his fingers incised. The whole of the party, however, ultimately recovered, and Count Zambeccari, in no way intimidated, continued to persevere in making ascents to a considerable height. In the year 1812, accompanied by Signor Bonagna, he ascended from Bologna. On coming down the balloon caught in some high trees and took fire; to avoid being burned they leaped out, where Count Zambeccari was killed, and his companion much injured.

In August 1808, Andreoli and Brioschi ascended at Padua, and rose rapidly to a considerable height. When the barometer had fallen to 15 inches, M. [Carlo] Brioschi felt a violent palpitation of the heart, and when it had reached 12 inches he sank into a state of torpor. M. Andreoli alone could observe the balloon, which rose till the mercury stood at 9 inches; he then found that he could not use his left arm. Soon after this, with the barometer at 8 inches, the balloon is said to have burst with a loud report, and then all came rapidly down together, with safety, near the place of Petrarch's Tomb. The accuracy of this statement has been questioned by [Thomas Forster] the author of *Aerial and Alpine Voyages*, who takes it for granted that the rapid escape of heated air would have caused not only a precipitate descent of the whole machine, but the death of the aeronauts. The only part of the account that I feel inclined to question would be that concerning the reading of the barometer, which gives an elevation of more than 30,000 feet. The resistance offered by the air does much in such cases, and it is not an inevitable result that everyone must be dashed to pieces. I have myself, under the pressure of an immediate necessity to save the land, fallen the last two miles in four minutes, holding to the valve line to ensure its opening to the full extent and the rapid escape of the gas, and though bruised have not been hurt severely. Mr. [John] Wise, the American aeronaut, has also twice descended to the earth with an exploded balloon. The canvas, torn and rent, acts as the mainsail of a ship, and the balloon gyrates through the air in falling. It is not by any means a situation to be coveted, but one, I should be understood to remark, not necessarily involving loss of life, even from so great an elevation as that of the Italian aeronauts. Increase of height accelerates the velocity of the descent, and much increases the hazard of the situation; but it is possible to fall and live. In one of my descents from Wolverhampton, the wind made it difficult and dangerous, and with our utmost efforts the balloon came roughly to the ground: it struck the earth and rebounded again and again, until a long tear became visible, which spread rapidly. The sides of the balloon stood out like wings, but the upper part remained, until finally a great rent passed up from neck to valve, when I fully expected all would drop down. But for some little time after this the great valve, with its heavy springs, remained fifty feet high in the air, whilst the whole balloon opened out in one immense sheet, and, kite-like, kept up perhaps for rather more than a minute, though it appeared to me a much longer time. It then gradually fell to the ground. That it did not fall more rapidly was due to the pressure of the atmosphere. We had a few bruises, but none of any importance, and were spared the general reversal of our effects which happened to Mr. Wise, who alighted with his car bottom upwards. If we therefore, in consideration of our own and other authenticated experiences, allow that the Italian aeronauts might have survived the catastrophe of their machine, and that the elevation they attained was nearly, if not quite, equal to that which they record, I may remark that the remainder of their statement bears comparison with the effect of rarefied atmosphere upon others. Thus, Signor Andreoli, of whose ascents there are frequent mention, and who was more inured probably to the higher regions, suffered less than Signor Brioschi, and observed the barometer after his companion became insensible. At 15 inches, Signor Brioschi found his respiration seriously affected. At 15 inches, I began to pant for breath. At 12 inches, Signor Brioschi became insensible—that is at about 23,000 feet above the earth, the same height to which Gay-Lussac attained without inconvenience. At 9 inches of the barometer, that is at 29,000 feet, Signor Andreoli, more seasoned than his companion, found only that he could not use his left arm, and was able to observe that the balloon was fully inflated. The balloon employed was doubtless one of Montgolfier's, filled with heated air, as such were principally in use in Italy. It ascended with great rapidity, and, unless they carried fire, must have cooled and descended within a very short period of time, without travelling far from the place of ascent. Leaving it an open question whether or not the barometer readings may have given too great altitudes, we may fairly suppose that Signor Andreoli, of whom frequent mention is made, had become enabled to support the greater rarity of the air, whilst his companion is shown to have yielded at an elevation less than that at which, when seasoned, I first became seriously affected. I am inclined to accept the statement as it is written, and the facts described are certainly in accordance with other experiences. It may be argued that Gay-Lussac felt no inconvenience at the height at which Signor Brioschi fainted; but the French philosopher ascended in a hydrogen balloon, slowly, to take note of the instrumental phenomena committed to his charge, and there is every reason to believe that Signors Brioschi and Andreoli ascended very rapidly to a great height, the sudden effect of which, as I have already said, is a shock which the system is unable to support.

In August 1811, Mr. [James] Sadler and Mr. Henry Beaufoy ascended from Hackney; they attained an elevation no higher than sufficient to view the landscape of the earth spread out beneath them like an open map, and were not therefore subject to the test of physiological sensations. The idea of unlimited freedom conveyed by the sense of floating in the invisible medium which surrounds the aeronaut; the total unconsciousness of movement, and the sudden sinking away of the earth and the people on it; the silence of the upper regions succeeding instantaneously to the shouts of the spectators and the noise and turmoil around the car, are among the first impressions which occurred to Mr. Beaufoy.

In 1812, Mr. Sadler ascended at Dublin to cross to Liverpool, but meeting with an adverse current, he resolved to descend into the sea. To escape from drowning, and effect the disablement of his balloon, he caused the crew of a ship to run her bowsprit through it, and then to take him on board. Mr. Sadler, junior, ascended from the Green Park, and with difficulty saved his life. Not only did the valve become frozen, but the net burst at the top, and the silken covering of the balloon began gradually protruding through it. To save himself from being precipitated to the earth, he tied the long silken neck of the balloon round his body. After being carried to a great height into the upper regions, and almost frozen with the cold, he came down at length near Gravesend.

In 1832, Dr. [Theodore] Foster ascended with Mr. Green from Chelmsford, with the idea of making some further observations on clouds, in addition to those already made on Alpine excursions; also to test by personal sensation the effect of the higher regions of the air upon the organs of hearing. They ascended slowly, and were for a time becalmed. "It was towards evening, and looking in the direction of Maldon River, and hovering over its marshy land, we saw," observes Dr. Foster, "what had evidently been a cumulus now subsiding into a stratus, or white evening mist, stretching in such a manner over the ground in its descent, that we at first took it for smoke. Higher up there were cumuli in the air, and uniform haze, and some warm clouds. The beauty and extent of prospect now increased. All earthly sounds ceased as soon as we had got above the breeze which swept above the surface of the ground, where in a region comparatively calm, and lighter than it was below, we were conscious of no motion whatever. I presently felt a slight movement, and heard the great buoyant balloon above us make a noise, as if touched by the wind. On adverting to the cause, we found that we had got into another current, which wafted us back again towards Chelmsford as we moved round with the oscillating machine . . . I remember, in crossing to France," continues Dr. Foster, "the first experience of a steamboat paddling across the level brine like a fish was a curious phenomenon, having before been only conveyed by sailing vessels. But this newborn Leviathan of the deep is nothing to this Pegasus of the air, neither is the sensation produced by a balloon in motion at all comparable to that of a balloon at rest."

The most remarkable ascent of the century was that fitted out by Robert Hollond, Esq., M.P. Mr. Green's balloon, afterwards known as the great *Nassau*, was employed for the expedition, and provided with every imaginable requisite, and provisions to last a fortnight, or longer if need be. On the afternoon of Monday, November 7, 1836, it left Vauxhall Gardens. The party consisted of Mr. Green and Mr. Hollond, the projector of the enterprise, accompanied by Mr. [Thomas] Monck Mason. It was one o'clock when they left the earth, and, in obedience to the prevailing current, were wafted gently along. By the fading light of the winter day they found themselves leaving land, and vertically placed above the breakers on the beach beneath. Throughout the night, in utter darkness, they voyaged for hours above a dense stratum of cloud, through breaks of which an occasional glimmer of light from the fires on the surface of the earth alone could penetrate by a partial glimpse. As morning dawned the aspect of the country they were traversing afforded them no knowledge of their bearing, and at ten minutes past five they gained their greatest elevation, and mounted to a height of 12,000 feet. At a quarter to six they were brought into full view of the sun, and presently descending, to rise again, enjoyed the spectacle of a sunrise above the clouds. As the sun gained power they anxiously endeavoured to gain some knowledge of the position they occupied above the earth, and, in ignorance of the speed with which they had been journeying, and of the distance traversed, began to surmise that they might already have passed the limit of that part of Europe where they might expect to find the accommodation and conveniences necessary for their comfort and the safety of the balloon. The large tracts of snow beneath them suggested the plains of Poland or the steppes of Russia; they therefore proposed to descend without delay, and, lowering the grapnel, came safely to earth, passing the gentle declivity of a wooded valley, and descending into the bosom of the trees which capped its summit. Bespeaking the assistance of people near, the balloon was speedily secured, and they learned that they had descended in the duchy of Nassau, about two leagues from the town of Weilburg. The journey had lasted eighteen hours, and was thus brought to a safe and agreeable termination. Mr. Monck Mason drew up an able account of the expedition, which he subsequently published in his *History of Aerostation*, a work to which I refer my readers who may feel interested in further particulars of the voyage.

Had I attempted a consecutive narrative of balloon ascents (instead of calling attention to those only which were important on account of their elevation), the names of Pilâtre de Rozier, the first aeronaut, and Blanchard, the first aerial voyager by profession, would have found greater prominence.

In the use of the balloon, distinction must be made between travelling for miles horizontally over a surface of country which is disclosed like a grand natural panorama to the eye of the voyager, and ascending perpendicularly to the greatest altitude within the capacity of the machine and the limits of human life. Vertical and horizontal explorations of the air have each a range of experiences of their own; the latter give rise to personal enjoyment chiefly, while the former add to our knowledge of hitherto unexplored territory.

For vertical motions only is the balloon manageable. With its capacity measured and weight determined, its ascending power can be calculated, and the aeronaut may nerve himself to brave the vicissitudes of a certain elevation, and, if inured to the work of observation, make every fresh ascent an epoch in discovery. To Mr. Green is due the employment of coal gas, which has long superseded the use of hydrogen. The filling a balloon, therefore, is no longer the tedious and uncertain operation it was formerly, extending sometimes over several days, but is performed with ease and certainty in a few hours and at a moderate cost. The comparatively easy management of a balloon so filled in the hands of a practised aeronaut, under whose guidance for a matter of *£ s. d.* one can sit securely and for an hour or two enjoy the delight of an aerial voyage within sight of earth, is one reason, I believe, why the balloon has gradually degenerated into an instrument of popular exhibition and passing amusement, so that its striking characteristics and important bearing are in danger of fading completely out of view.

To guide the balloon in any horizontal direction appears now as far from practicable as it ever has been. We start from a given point to go where chance directs. The compass we carry with us, not that we may steer our course along a given route, but trace by it the erratic and ungoverned movements of the machine that carries us. We traverse perhaps the segment of a huge circle, the line of our path in space. We proceed and return, advance onward, now gently, now with velocity. We sit in the car without the slightest knowledge of the earth's landscape hidden beneath the vapours of the air. The voyage itself is to last many hours, if all things should be favourable. Where, let us ask, is the practical advantage of such a machine? To what use can it be converted? Are we wrong in supposing it to be a first principle which requires yet to be engrafted into some mechanism which shall be more subordinate to the requirements of life?

It is not to be supposed that additional frequency of respiration in an attenuated air makes amends for the want of oxygen. Those who have felt the continued dryness of the throat, which is parched so that to swallow is painful, are sensible to the contrary; but the death it produces is painless, and asphyxia steals away the life of the human being as he moves above, suspended in mid-air, as stealthily as cold does that of the mountain traveller, who, benumbed and insensible to suffering, yields to the lethargy of approaching sleep, and reposes to wake no more. These two powers rule respectively the upper regions of the atmosphere, whether we seek to approach them by vertical ascent or by the steepest mountains, and the element we live in warns back the adventurous traveller to the limits appointed to human life and physical exertion.

Let us take the balloon as we now find it, and apply it to the uses of vertical ascent; let us make it subservient to the purposes of war, an instrument of legitimate strategy; or employ it to ascend to the verge of our lower atmosphere; and as it is, the balloon will claim its place among the most important of human inventions, even if it remain an isolated power, and should never become engrafted as the ruling principle of the mechanism we have yet to seek.

The balloon considered as an instrument for vertical exploration, presents itself to us under a variety of aspects, each one of which is fertile in suggestions. Regarding the atmosphere as the great laboratory of changes which contain the germ of future discoveries to belong respectively as they unfold to the chemist and the meteorologist, the physical relation to animal life of different heights; the form of death which at certain elevations waits to accomplish its destruction; the effect of diminished pressure upon individuals similarly placed; the comparison of mountain ascents with the experiences of aeronauts, are some of the questions which suggest themselves, and faintly indicate inquiries which naturally ally themselves to the course of balloon experiments. Sufficiently varied and important, they will be seen to rank the balloon as a valuable aid to the uses of philosophy, and rescue it from the impending degradation of continuing a toy, fit only to be exhibited, or to administer to the pleasures of the curious and lovers of adventure.

We can also make use of it to determine the proportions of the gaseous elements we breathe. Do not the waves of the aerial ocean contain, within their nameless shores, a thousand discoveries destined to be developed in the hands of chemists, meteorologists, and physicists? Have we not to study the manner in which the vital functions are accomplished at different heights, and the way in which death takes possession of the creatures whom we transport to these remote regions? Have we not to compare the different effects of the diminution of pressure on individuals placed in identical condition in the car of the same balloon?

When the balloon was invented, the great [Antoine-Laurent] Lavoisier was charged by the Academy of Sciences to draw up a report in order to estimate the value of this unexpected discovery. After having minutely described the ascents at which he was present, the illustrious chemist stopped, appalled in some measure at the multitude of the problems the balloon would help to solve, and the series of uses of which it seemed susceptible. I shall imitate his reserve; for it seems unnecessary to justify further the attempt to make the balloon a philosophical instrument, instead of an object of exhibition, or a vehicle for carrying into the higher regions of the air excursionists desirous of excitement, mere seekers after adventure.

Chapter I

The first scientific ascents in England

There are no frontiers in the reign of thought, and the conquests of the human mind belong to all the world; yet each civilized nation is called upon to give its contingent to the great work of the study of nature, and to choose those branches which are most suited to its genius.

France has given the balloon to the world, but her work is still incomplete, and the conquest of Charles and Montgolfier remains undeveloped. It is not, however, my intention to describe the attempts which have been made to this end, or discuss the value of the balloon as a first step towards the solution of the problem of aerial locomotion; I desire only to describe the principal results of my own aeronautical excursions, after briefly alluding to the observations of my predecessors in this field of inquiry.

The first persons in England who devoted themselves to aerial navigation were foreigners. The philosopher Tiberius Cavallo and the diplomatist Vincent Lunardi were both Italians. But from the time when Lunardi inaugurated balloon ascents to the present day, it may be truly said that balloons have remained popular with us; not only have noblemen and gentlemen shown a taste for aerial journeys, but men of science have followed up with avidity the great experiments made on the Continent, and several attempts have been made in England, both by free and captive balloons, to study systematically the phenomena of the atmosphere.

In 1838 and 1850, Mr. [George] Rush ascended several times with Mr. Green, and made some observations mainly on humidity. Public attention was aroused to a certain extent, but the ascents were chiefly known from an incident which occurred at the end of one of them. The balloon descended in the sea near Sheerness, and the car was dragged through the water with considerable rapidity; the balloon acting as a kite. Mr. Green therefore threw out the grapnel, which caught in a sunken wreck, and detained the balloon till a boat came up and secured the voyagers. A volley of musketry was fired into the balloon to admit of the escape of the gas, and it was ultimately secured.

Soon after the discovery of the balloon, a desire arose for experiments in the higher regions of the air. The first experiments, as I have previously stated, were made at St Petersburg, by command of the Emperor of Russia, by Mr. Robertson, in the years 1803 and 1804, but no important results were obtained.

In the year 1804 two experiments were made at Paris: the first on August 31, by Gay-Lussac and [Jean-Baptiste] Biot. These gentlemen ascended to the height of 13,000 feet, but did not commence their observations till they were 7,000 feet high. Their experiments in magnetism, electricity, or galvanism, gave results identical with those made on the earth—a source of much disappointment to everyone.

It was then supposed that they had not ascended high

enough, and Gay-Lussac resolved to go alone, with the view of reaching a greater elevation. This he succeeded in doing on September 15 following, when he reached a height of 23,000 feet, and found a decline of temperature from 82° to 15°; almost confirming the theory of a decline of temperature of 1° in 300 feet of elevation. The sky was very blue, and the air was found to be very dry. A magnet took a longer time to vibrate than on the earth. He filled two bottles with air from the higher regions, which on analysis was found to be in its component parts the same as the lower air.

Two years after this, the Astronomer Royal of Naples, Carlo Brioschi, wished to ascend higher than Gay-Lussac, but this he was unable to do in consequence of the balloon bursting. After this no attempt was made till the year 1843, when the British Association appointed a committee and voted a sum of money for experiments by means of captive balloons. Several committees were subsequently appointed, and out of the limited resources of the Association considerable sums of money were granted for experiments by means of balloons; but no good results were obtained. This want of success ought neither to discourage nor astonish us; captive ascents, though easy enough when directed by experienced aeronauts with proper appliances, present inextricable difficulties to novices unaccustomed to the disappointments of aerial navigation.

In the year 1850 Messrs [Jacques Alexandre] Bixio and [Augustin] Barral conceived the project of ascending to a height of 30,000 to 40,000 feet, in order to study the many atmospheric phenomena as yet imperfectly known. On June 29 in that year, a balloon was filled in the garden of the Observatory at Paris with *pure hydrogen gas*. The weather was bad—a torrent of rain fell;

Messrs Bixio and Barral, and the aeronaut, placed themselves in the car without testing the ascending power of the balloon, and darted into the air like an arrow, as described by the spectators, so that in two minutes they were lost in the clouds. At a height of 5,000 feet the gas in the balloon expanded with great force against the netting, which proved to be too small. The balloon became full, and descending upon the voyagers covered them completely as they were seated in the car, which unfortunately was suspended by cords much too short. In this difficult situation, one of them, in his efforts to disengage the cord from the valve, made an opening in the lower part of the balloon, from which the gas escaping at the height of their heads, occasioned them continued illness. Then they found that the balloon was torn and they were falling fast. They threw away everything they could, and came to the earth in a vineyard, having left it only forty-seven minutes previously. A mass of clouds 9,000 feet in thickness was passed through. The decrease of temperature up to 19,000 feet, the highest point reached, seemed to confirm the results obtained by Gay-Lussac in 1804.

In the following month, on July 27, the filling of the balloon was commenced early in the morning. It proved to be a long operation, occupying till nearly two o'clock: then heavy rain fell, the sky became overcast, and it was after four when they left the earth. They soon entered a cloud at 7,000 or 8,000 feet, which proved to be fully 15,000 feet in thickness; they never, however, reached its highest point, for when at 4:50 a.m. the height of 23,000 feet was reached, they began to descend, owing to a tear which was then found in the balloon. After vainly attempting to check this involuntary descent, they reached the earth at 5:30 a.m.

On approaching the limit of this cloud of 15,000 feet in thickness, the blue sky was seen through an opening in the surrounding vapour. The polariscope, when directed towards this point, showed an intense polarization, but when directed to the side, away from the opening, there was no polarization.

An interesting optical phenomenon was observed in this ascent. When near their highest point, the bed of clouds which covered the balloon having become less dense, the two observers saw the sun dim and quite white, and also at the same time a second sun reflected as from a sheet of water, probably formed by the reflection of luminous rays on horizontal sides of crystal ice floating in the clouds.

The most extraordinary and unexpected result, however, observed in this ascent was the great change of temperature. At the height of about 19,000 feet the temperature was 15°, but in the next 2,000 feet it fell to minus 39°. This wonderful change was experienced in the clouds. What, we may ask, can the constituents of such a cloud then be? In this voyage a height short of Gay-Lussac's by 50 feet was reached, but a temperature lower by 54° was recorded, and the clothes of the observers were covered with fine needles of ice. From this time until quite recently no ascents have been made in France in the cause of science.

In the year 1852 Mr. [John] Welsh, of the Kew Observatory, made, under the auspices of the British Association, four ascents in the great *Nassau* balloon, with the veteran aeronaut Mr. Green, who had then an experience derived from several hundred ascents.

In August, October, and November he reached the respective heights of 19,500, 19,100, 12,640, and 22,930 feet, and in each ascent made a valuable series of observations.

The facts recorded by Gay-Lussac, relative to the decline of temperature with increase of elevation, appeared to confirm the law which had been derived from observations made on mountain-sides, viz. a decrease of 1° for every increase of 300 feet of elevation; and the deductions of Mr. Welsh from his experiments tended to the confirmation of the same law, with some modifications.

The results of Welsh's observations were published in the *Philosophical Transactions of the Royal Society* for the year 1853, and afterwards in the *Bulletin Géographique de Dr. Petermann* for 1856.

When these ascents were made, they excited the greatest public interest. I watched Mr. Welsh's fourth ascent throughout, from the roof of the Royal Observatory at Greenwich, with a good telescope. The day was fine and the air clear, and I was surprised at the facility with which I could follow every movement of the balloon, from its departure to its descent. During the whole time that the balloon was in the air, and while it traversed a course of fifty-seven miles in the direction E.S.E., I never lost sight of it for a moment. I saw it rise from Vauxhall at 2:22 p.m., and descend at 3:40 p.m., at a place which I afterwards learned was near Folkestone. It was this circumstance which notably influenced me in my desire for balloon observations, and which led me to believe in the possibility of combining terrestrial observations with those made in the balloon, and thereby determining the height of the balloon at different times, independently of observations made in the car. But in my own ascents I never was able to organize, to my satisfaction, the telescopic observations of the balloon from the earth, so as to verify the heights determined from my own observations.

This, however, was not the first time aerial physics had engaged my attention. A taste for these studies was first developed during my residence in Ireland in the years 1829 and 1830. In these years I was often enveloped in fog for entire weeks, first on the mountain Bencor, in Galway, and afterwards upon the summit of the Keeper Mountain, near Limerick. At this time I was engaged on the principal triangulation of the Trigonometrical Survey of Ireland, and in the performance of my duty I was often compelled to remain, sometimes for long periods, above, or enveloped in cloud. I was thus led to study the colours of the sky, the delicate tints of the clouds, the motion of opaque masses, the forms of the crystals of snow. On leaving the Survey, and entering the Observatory of Cambridge, and afterwards that of Greenwich, my taste did not change. Often between astronomical observations I have watched with great interest the forms of the clouds, and often, when a barrier of cloud has suddenly concealed the stars from view, I have wished to know the cause of their rapid formation, and the processes in action around them.

The illness of Mr. Welsh interrupted his series of experiments, and scientific ascents ceased to occupy public attention. But the British Association did not lose their interest in aerial experiments, and Colonel [William Henry] Sykes, M.P. for Aberdeen, again brought the subject before the meeting of the British Association at Leeds in 1858, and obtained the appointment of an influential committee. The resources of the Association, composed exclusively of the contributions of its members, are devoted mainly to taking the initiative in important and hitherto unexplored departments of science, and out of these limited means the necessary grants for these scientific balloon ascents were made, the chief expenses being the hire of the balloon, the payment of the aeronaut for its management, and the cost of the gas. Several of the members of the committee had already made balloon ascents with Mr. Green. They were, therefore, well able to appreciate the importance of observations made in and above the clouds. It was at first arranged that Mr. Green should direct the ascents, and that the observations should be taken by young men. Mr. Green, who was born in 1785 the same year as the introduction of balloons into England, was then seventy-four years of age.

I gave two young observers all the instructions I could in respect to the observations to be taken, and explained to them all the precautions that a long life devoted to observations suggested to me. On August 15, 1859, the members of the committee met at Wolverhampton, in order to assist at the first departure of the balloon. This town was selected on account of its central position. It was subsequently the point of departure of some of my most successful expeditions.

The weather was fine when the filling of the great *Nassau* balloon was begun; but the wind arose, and many accidents happened which prevented the filling of the balloon taking place, so that the ascent was deferred till August 16. The committee was again at its post on this day, but, as it proved, only to see an aerial shipwreck. When many thousands of feet of gas had been introduced into the balloon, the wind arose and blew it with such violence that it was torn, and all the gas escaped.

Mr. Green, having examined the injury, said it would take many days to repair, and as the meeting of the Association was approaching, it was resolved to defer the experiment. Such accidents would be impossible, or at least of extremely rare occurrence, if a less barbarous mode of inflating balloons than filling them slowly in the open air were adopted.

Mr. Green was greatly distressed at this accident, which was due to no fault of his; for it was attributable to the interruption of a series of experiments which, he calculated, would have placed aerial navigation in its proper place, and raised it from the inferior position in which he found it. Having had, he said, all his life to contend with similar difficulties at places of amusement only, he was more than anyone else aware of the importance of experiments made under irreproachable conditions, and placed under the patronage of learned men; and he wished to close his career under such circumstances.

The career of Green began in 1821, the year of the coronation of George IV; it continued for thirty-six years, during which he made nearly 1,400 ascents. Three times he crossed the sea; twice he fell into it. He obtained a large experience, and his accounts are worthy of all confidence; but, unfortunately, his education was not sufficiently good to make him a competent observer in the higher regions of the atmosphere. However, he improved the general management of balloons in many particulars—his guide rope in aerial navigation, particularly of use in crossing seas, and the introduction of carburetted gas in the place of hydrogen, are worthy of mention. He died in the year 1870, in his eighty-sixth year.

The Balloon Committee, though discouraged by these frequent delays, resolved to organize four ascents from Wolverhampton. It was decided that they should be to the height of four or five miles, in order to verify the facts announced by Gay-Lussac and Messrs Bixio and Barral; but on inquiry it was found that no balloon that would contain a sufficient quantity of gas to enable an observer to ascend so high was to be obtained in England. The largest, it was understood, was the *Royal Cremorne*, which would hold nearly 50,000 feet. This balloon the committee therefore obtained, and Mr. [Thomas] Lythgoe, who had made nearly one hundred ascents, principally from Cremorne, was employed as aeronaut. Ballooning had been for many years pursued only as a trade, and there was no choice whatever either of balloons or aeronauts. Notwithstanding the desire which I had always felt for observations at high altitudes, I had decided not to take the observations myself, but only to give all necessary instructions in the use of instruments and precautions necessary to be taken.

As the gentleman who first engaged to be the observer declined, the observations were entrusted to Mr. [Henry Charles] Criswick, assistant at the Observatory at Greenwich, who alone was to accompany the aeronaut. The space within the boundary of the Gas Works was selected for inflating the balloon. Before the hour of the ascent, the members of the committee, with Lord Wrottesley and Mr. W. [Sir William] Fairbairn, the President of the British Association, were on the ground.

At 1:04 p.m. the balloon ascended slowly and steadily. After remaining nearly stationary for a few minutes sand was thrown out, and the height of one mile was reached; in thirteen minutes it passed out of sight; but little more than a mile had been reached when the balloon descended from sheer inanition. It proved to be full of minute holes, and was quite useless, as were the observations made, which contradicted themselves. The disappointment was great. Arrangements had been made for meteorological observations every few minutes, at thirty different places. This check to the proceedings was very serious, and naturally disgusted many with aeronautical experiences. Colonel Sykes and the committee were bitterly disappointed, but met in consultation at Wrottesley Hall. Mr. Lithgoe admitted that the balloon had been in use thirty years, and was worn out; he advised application to be made to Mr. [Henry] Coxwell for the use of his *Mars* balloon.

I must ask pardon of the reader for entering into all these details, but they show the greatness of the difficulties with which such investigations are too often surrounded. One would have believed that the real difficulties would have been met with in the air, but, on the contrary, the greatest difficulties had to be overcome on the earth.

The *Mars* was found to be injured. Several tailors were set to repair it, but it was found that their combined labour could not effect the reparation in less than several days, and even then Mr. Coxwell said he could not pledge himself to make a safe ascent; he offered, however, to construct a new balloon, larger than any previously made. It was in the car of this balloon that by far the greater number of my experiments were subsequently made.

Chapter II

My first ascent—Wolverhampton

July 17, 1862

Notwithstanding all these accumulated difficulties and the efforts I had been obliged to make to overcome them, I found that in spite of myself I was pledged both in the eyes of the public and the British Association to produce some results in return for the money expended. I therefore offered to make the observations myself. The three or four months which elapsed between the abortive attempt of the *Mars* and my first ascent were devoted to preparatory studies and experiments; for I was occupied with the construction and management of the apparatus which I intended to take with me. I also accustomed myself to the use and manipulation of the instruments in a limited space, and considered how best to group them on a board such as would have to serve me for a table in the car of the great balloon; so that when the day for the ascent came, I was able to imagine that I was not making my aerial debut.

In spite of the experience which I had of observations on the earth, and in spite of the time which I had devoted to this first ascent, I had neglected a great number of useful precautions and encumbered myself with some superfluous apparatus; in short, I was able to perfect without cessation my apparatus in every successive ascent. I hope that the experience which I have acquired, sometimes to my cost, will show how much those philosophers are in error who think that observations in the higher regions can be made well enough by the first observer that comes.

The novelty of the situation, the rapidity with which all the observations must be made, and the smallness of the space at command, require that the observer should have previously had considerable practice in the use of the instruments under all circumstances. I may mention also that I experienced great anxiety when I reflected that at every instant I might be failing to observe very important phenomena, and that I was excessively fatigued by the extraordinary attention to which I found myself condemned by the fear of not being ready when the moment came to observe a phenomenon which perhaps no human eye had contemplated before.

The objects to which the Committee of the British Association resolved to devote their principal attention were, primarily:—To determine the temperature of the air and its hygrometrical states, at different elevations, to as great a height as possible; to determine the rate of decrease of temperature with increase of elevation, and to ascertain whether the results obtained by observations on mountain-sides, viz. a lowering of temperature of 1° degree for every increase of elevation of 300 feet, be true or not; also to investigate the distribution of the water in the invisible shape of vapours, in the air below the clouds, in the clouds, and above them at different elevations. Secondarily:—

1. To determine the temperature of the dew point by [Jean Frederic] Daniell's dew point hygrometer, by [Henri Victor] Regnault's condensing hygrometer, and by dry and wet bulb thermometers as ordinarily used, as well as when under the influence of the aspirator (so that considerable volumes of air were made to pass over both their bulbs) at different elevations, as high as possible, but particularly up to those heights where man may be resident, or where troops may be located (as in the highlands and plains of India), with the view of ascertaining what confidence may be placed in the use of the dry and wet bulb thermometers at those elevations, by comparison with the results as found from them, and with those found directly by Daniell's and Regnault's hygrometers; also to compare the results as found from the two hygrometers.
2. To compare the readings of an aneroid barometer with those of a mercurial barometer up to five miles.
3. To examine the electrical condition of the air at different heights.
4. To determine the oxygenic condition of the atmosphere by means of ozone papers.
5. To determine whether the horizontal intensity of the earth's magnetism was less or greater with elevation, by the time of vibration of a magnet.
6. To determine whether the solar spectrum, when viewed from the earth, and far above it, exhibited any difference, and whether there were a greater or less number of dark lines crossing it, particularly near sunset.
7. To collect air at different elevations.

8. To note the height and kind of clouds and their density and thickness.
9. To determine the rate and direction of different currents in the atmosphere.
10. To make observations on sound.
11. To make observations on solar radiation at different heights.
12. To determine the actinic effects of the sun at different elevations by means of [John] Herschel's actinometer.
13. To note atmospherical phenomena in general, and to make general observations.

Everyone knows that *the pressure of the atmosphere* is measured by means of the barometer. A column of air extending to its limit of the same area as the barometer tube is balanced by the column of mercury in the tube; and if we weigh the mercury, we know the weight or pressure of the column of atmosphere upon that area. If the area of the barometer tube be one square inch, then this would tell us the pressure of the atmosphere on one square inch. The length of a column of mercury thus balanced by the atmosphere, near the level of the sea, is usually about 30 inches, and if this be weighed it will be found to be nearly 15lbs: therefore the atmospheric pressure on every square inch of surface is about 15lbs—just one-half as many pounds as the number of inches which expresses the height of the column of mercury.

Now, in ascending into the air, part of the atmosphere is below, and part above: the barometer therefore has to balance that which is above only, and will therefore read less.

At the height of three miles and three-quarters, the barometer will read about 15 inches: there is therefore as much atmosphere above this point as there is below, and the pressure on a square inch is 7.5lbs.

At a height of between five and six miles from the earth, the barometer reading will be about 10 inches: one-third of the whole atmosphere is then above, and two-thirds beneath; and the pressure on a square inch is reduced to 5lbs.

The reading of the barometer varies with the altitude at which it is observed, and indicates by its increasing or decreasing readings corresponding changes in the pressure of the atmosphere.

At the height of	1	the barometer reading is	24.7 inches
At the height of	2	the barometer reading is	20.3 inches
At the height of	3	the barometer reading is	16.7 inches
At the height of	4	the barometer reading is	13.7 inches
At the height of	5	the barometer reading is	11.3 inches
At the height of	10	the barometer reading is	4.2 inches
At the height of	15	the barometer reading is	1.6 inches
At the height of	20	the barometer reading is	1.0 inches

By the reading of the barometer in the balloon, the distance from the earth is known; and if the balloon be situated above clouds, or in a fog, the reading of the barometer indicates the near approach of the earth, and acts as a warning to the occupants of the car to prepare accordingly. In addition to this temporary use, the readings combined with those of temperature enable us to calculate the height of the balloon at every instant at which such readings have been taken.

The temperature of the dew point also deserves a few explanatory words.

There is always mixed with the air a certain quantity of water, in the invisible shape of vapour, sometimes more, sometimes less; but there is a definite amount which saturates the air at every temperature, though this amount varies considerably with different temperatures.

A cubic foot of air at the temperature of—

30°	is saturated with	2	grains of	vapour of water.
49°	is saturated with	4	grains of	vapour of water
70°	is saturated with	8	grains of	vapour of water
92.5°	is saturated with	16	grains of	vapour of water

The capacity of air for moisture therefore doubles for every increase of temperature of about 20 degrees.

The temperature of the dew point is the temperature to which air must be reduced in order to become saturated by the water then mixed with it; or it is that temperature to which any substance, such as the bright bulb of a hygrometer, must be reduced before any of the aqueous vapour present will be deposited as water, and become visible as dew. The temperature at which this first bedewing or dulling of bright surfaces takes place is the temperature of the dew point. For instance, I have already said that two grains of water saturate a cubic foot of air at 30°: if, therefore, the temperature of the air be 40°, and there be two grains of moisture in a cubic foot of air, then, if the bulb of the hygrometer be reduced to 30°, a ring of dew will appear on it, caused by the deposition of the water in the air.

The determination of the dew point at once tells us therefore the amount of water present, and, combined with the temperature, enables us to determine the hygrometrical state of the atmosphere.

If the air be saturated with moisture, the temperature of the air and that of the dew point are alike; if it be not saturated, the temperature of the dew point is lower than that of the atmosphere; if there be a great difference between the two temperatures, the air is dry; and if this happen when the temperature is low, there is very little water present in the air.

By the careful simultaneous readings of two thermometers, one with a moistened bulb and the other dry, or by the use of a Daniell's or Regnault's hygrometer, the amount of water present in the air in the invisible shape of vapour can be determined, as well as the temperature of the dew point and the degree of humidity.

The degree of humidity of the air expresses the ratio between the amount of water then mixed with it and the greatest amount it could hold in solution at its then temperature, upon the supposition that the saturated air is represented by 100, and air deprived of all moisture by 0. Thus: Suppose the water present to be one-half of the quantity that could be present, the degree of humidity in this case will be 50. If the air were at the temperature of 30°, and there were two grains of moisture in the air, it would be saturated, and the degree of humidity would be 100. If there were one grain, that is one-half of the whole quantity that could be present, the air would be one-half saturated, and the degree of humidity would be represented by 50.

At	49°	with	4	grains of moisture	The air is saturated, and the degree of humidity is 100.
At	70°	with	8	grains of moisture	
At	92.5°	with	16	grains of moisture	
But at	49°	with	2	grains of moisture	The air is one half saturated, and the degree of humidity is 50.
At	70°	with	4	grains of moisture	
At	92.5°	with	8	grains of moisture	

The thermometers employed in the observations were exceedingly sensitive; the bulbs, long and cylindrical, being almost three-tenths of an inch in length. The graduations, which extended to minus 40°, were all made on ivory scales. These thermometers, on being removed from a room heated 20° above that of the surrounding air, acquired the temperature within half a degree in about ten or twelve seconds. They were so sensitive that no correction was necessary for sluggishness; and this was proved to be the case by the near agreement of the readings at the same height in the ascending and descending curves, in cases when there was no reason to suppose there had been any change of temperature at the same height within the interval between the two series of observations.

I had two pairs of dry and wet bulb thermometers; one pair similar to those ordinarily used, the bulbs being protected from the direct rays of the sun by a highly polished silver shade, in the form of a frustum of a cone, open at top and bottom, and a cistern fixed near to them for the supply of water to the wet-bulb thermometer.

The second pair were arranged for the employment of the aspirator, the object of which was to induce at will a current of air across the bulbs, which, being highly sensitive, would almost instantaneously record the temperature of the air so set in motion. In this arrangement the thermometers were enclosed in silver tubes placed side by side, connected together at top by a cross tube, and both protected by a shade, as in ordinary use. In the left-hand tube belonging to the dry-bulb an opening was provided. By means of the aspirator a current of air was drawn in at this opening, which, traversing round the tubes, passed away into the aspirator. Thus the temperature of the air in motion against the bulbs could be determined at pleasure with the utmost nicety.

Regnault's condensing hygrometer was made with two thermometers, and as described by Regnault himself. The scales were made of ivory, and the thermometers fitted to the cups with cork, ready for packing up at short notice.

All the instruments were attached to the table with strings, which could be cut immediately, or they merely rested on stands which were screwed to the table. This table was fixed across the car, and tied there by strong cord. On approaching the earth, all the instruments were rapidly removed and placed, anyhow, in a basket, furnished with a number of soft cushions to cover them in layers, so that they were not broken by the shock on coming in contact with the earth. When more than two or three persons were in the car, besides myself, the arrangement of the instruments was different, and they were fewer in number.

As such ascents (when several were in the car) of necessity could not be of extreme heights, and as it was found in the high ascents that the aneroid read at all times very nearly the same as the mercurial barometer, the same aneroid which had thus been tested was alone used for the determination of elevation, and the mercurial barometer was therefore not taken up.

It had also always been found that the dry and wet bulb thermometers, whether aspirated or not, read alike; the use of those under the influence of the aspirator was therefore dispensed with; and as in point of fact one thermometer and one bright surface are all that Regnault's hygrometer needs to determine the temperature of the deposit of dew, one of the thermometers only was used.

By these alterations I was enabled to conveniently place all the necessary instruments in a much smaller space; and ultimately, in my low ascents, I managed to place them all on a board, projecting beyond the side of the car, which had the double advantage of allowing the air to play more freely about them, and leaving the aeronaut more room. There was also a third arrangement adopted, viz. that for night ascents. The inconvenience of reading instruments at night necessitates the use of even a smaller number. In such experiments I have usually confined myself to the determination of the temperature and humidity of the air at different elevations by the use of the dry and wet thermometers solely.

In the night ascents I took with me a well-made Davy safety lamp, having previously tested it by plunging it lighted into gas proceeding from a pipe. I also took the lamp up on a day ascent, and found it could be used in a balloon car with perfect ease. By its use I was therefore able to read the instruments at night, though less quickly than in daylight. I used the same framework, placed outside the car as before, so that I stood with my back towards the aeronaut to whom the management of the balloon was entrusted. At night I also used to place a padded cushion, fitted into the frame, with padded sides, and in this I placed the watch, barometer, pencils, etc.

I have been thus particular in describing my arrangements, as they are the result of much thought and care, based upon experience.

At times I have taken up other instruments, such as the spectroscope, ozone tests, an actinometer, etc.; and this I was enabled to do when I found I could dispense with all the aspirating apparatus and some of the other instruments which were thought to be necessary at first: these I do not think I need particularize. The great principle to be attended to in the arrangement of the table is to fix everything by nuts, screws, or strings, and to place the instruments in such positions that they can be read with rapidity and ease, and removed in a very short time into a wadded case, so that they are not broken by the concussion.

On June 30, 1862, Mr. Coxwell brought his new balloon to Wolverhampton; it was not made of silk, but of American cloth, a material possessed of a great strength. Its capacity was 90,000 cubic feet, exceeding in size that of the famous *Nassau* balloon. Misfortune again followed the attempts of the committee; for, notwithstanding frequent uncomfortable gusts of wind, the inflation of the balloon was proceeded with, and after three hours about 60,000 feet of gas had passed in. At this time the wind arose, and great apprehensions were felt for the safety of the balloon, so that the supply of gas had to be cut off. The fierceness of the wind increased, and the balloon split upwards to the first cross seam, and taking the course of the seam, the rent ran almost round the balloon at its widest part. So much injury was done that it took more than a week to repair it, although many persons were employed on the work.

The directors of the Gas Company, and their engineer, Mr. Proud, very kindly consented to make, and to store away, some light gas, which we could not otherwise have procured. It is known that the products of the distillation of coal in a closed retort are richest in illuminating power at the commencement of the operation, and that their value diminishes as the distillation proceeds. The products of the last distillation are composed of a light gas, of weak illuminating power, but most suitable for balloon ascents. These last products were put into a special gasometer, and it is due to this circumstance that I was enabled to make the extreme high ascents, which would have been quite impossible if the Company had not placed a gasometer at our disposal.

After the balloon was repaired, a week's bad weather followed, and July 17 was the last day my engagements permitted me to remain at Wolverhampton. The filling of the balloon began at five o'clock in the morning, in the presence of Lord Wrottesley. As it proceeded, the weather increased in badness;

and if it had not been for the already great loss of time and the continued postponement of the ascent that would otherwise have taken place, we should not have set at defiance the terrible W.S.W. wind, which was blowing without interruption. Very great difficulties were experienced in the inflation, and it seemed as if the operation would never be completed. The movements of the balloon were so great and so rapid, that it was impossible to fix a single instrument in its position before quitting the earth, and the state of affairs was by no means cheering to a novice who had never before put his foot in the car of a balloon. When Mr. Coxwell made up his mind, at 9:42 a.m., to let go, the balloon, which had been so impatient to be free, did not rise, but moved horizontally on the ground for some distance, dragging the car on its side; which movement would have been fatal had there been any chimney or lofty building in the way.

We left the earth at about 9:43 a.m., and at 9:49 a.m. reached the clouds at an elevation of 4,467 feet. Rising still higher, at 9:51 a.m., with an elevation of 5,802 feet, we passed out of this stratum of cloud, but again became enveloped in a cumulo-stratus at the height of 7,980 feet. The sun shone brightly upon us at 9:55 a.m., and caused the gas to expand and the balloon itself to assume the shape of a perfect globe. A most magnificent view now presented itself, but, unfortunately, I was not able to devote any time to note its peculiarities and its beauty, as I was still arranging my instruments in the positions they were to occupy, and we had reached a height exceeding 10,000 feet before all the instruments were in working order. The clouds at this time (10:02 a.m.) were very beautiful, and at 10:03 a.m., at an elevation of 12,709 feet, a band of music was heard. At 10:04 a.m. the earth became visible through breaks in the clouds. At 16,914 feet the clouds were far below us, both cumulus and stratus, however, at a distance appearing to be at the same height as ourselves, the sky above us being perfectly cloudless and of an intense Prussian blue.

At starting, the temperature of the air was 59°, and the dew point 55°; at 4,000 feet it was 45°, dew point 33°, and it descended to 26° at 10,000 feet, dew point 19°; and then there was no variation of temperature between this height and 13,000 feet. During the time of passing through this space an addition was made to our clothing, as we felt certain we should experience a temperature below zero before we reached the height of five miles; but, to my surprise, at the height of 15,500 feet the temperature as shown by all the sensitive instruments was 31°, with a dew point of 25°, and at each successive reading up to 19,500 feet, the temperature increased, and was 42° at this height, with dew point at 24°. We had both thrown off all extra clothing. Within two minutes after this time, when we had fallen somewhat, the temperature again began to decrease with extraordinary rapidity to 16°, or 27° less than it was twenty-six minutes previously.

At the height of 18,844 feet, eighteen vibrations of a horizontal magnet occupied 26.8^{s}, and at the same height my pulse beat at the rate of 100 pulsations per minute. At 19,415 feet palpitation of the heart became perceptible, the beating of the chronometer seemed very loud, and my breathing became affected. At 19,435 feet my pulse had accelerated, and it was with increasing difficulty that I could read the instruments; the palpitation of the heart was very perceptible. The hands and lips assumed a dark bluish colour, but not the face. At 20,238 feet, twenty-eight vibrations of a horizontal magnet occupied 43^{s}. At 21,792 feet I experienced a feeling analogous to sea-sickness, though there was neither pitching nor rolling in the balloon; and through this illness I was unable to watch the instruments long enough to lower the temperature to get a deposit of dew. The sky at this elevation was of a very deep blue colour, and the clouds were far below us. At 22,357 feet I endeavoured to make the magnet vibrate, but could not; it moved through arcs of about 20°, and then settled suddenly.

Our descent began a little after 11 a.m., Mr. Coxwell experiencing considerable uneasiness at our too close vicinity to the Wash. We came down quickly, passing from a height of 16,300 feet to one of 12,400 feet between 11:37 a.m. and 11:38 a.m.; at this elevation we entered into a dense cloud which proved to be no less than 8,000 feet in thickness, and whilst passing through this the balloon was invisible from the car. From the rapidity of the descent the balloon assumed the shape of a parachute; and though Mr. Coxwell had reserved a large amount of ballast, which he discharged as quickly as possible, we collected so much weight by the condensation of the immense amount of vapour through which we passed, that notwithstanding all his exertions we came to the earth with a very considerable shock, which broke nearly all the instruments. All the sand was discharged when we were at a considerable elevation. The amount we had at our disposal at the height of five miles was fully 500lbs; this seemed to be more than ample, and, when compared with that retained by Gay-Lussac, viz. 33lbs, and by Rush and Green, when the barometer reading was eleven inches, viz. 70lbs., seemed indeed to be more than we could possibly need; yet it proved to be insufficient.

The descent took place at Langham, near Oakham, in Rutlandshire.

Chapter III

Ascents from Wolverhampton

August 18, 1862

The weather on this day was favourable; there was but little wind from the N.E. By noon the balloon was nearly inflated. As it merely swayed in the light wind, the instruments were fixed before starting, and at 1:02 p.m. and 38 seconds the spring catch was pulled, when for a moment the balloon remained motionless, and then rose slowly and steadily. In about ten minutes we passed into a magnificent cumulus cloud, and emerged from it into a clear space, with a beautiful deep blue sky, dotted with cirri, leaving beneath us an exceedingly beautiful mass of cumulus clouds, displaying a variety of magnificent lights and shades. Our direction was towards Birmingham, which came into view about 1:15 p.m.

When at the height of nearly 12,000 feet, with the temperature at 38°, or 30° less than on the ground, and the dew point at 26°, the valve was opened, and we descended to a little above 3,000 feet. The view became most glorious: very fine cumulus clouds were situated far below, and plains of clouds were visible to a great distance. Wolverhampton, beneath us, was sharp and well defined, appearing like a model. The clouds during this ascent were remarkable for their supreme beauty, presenting at times mountain scenes of endless variety and grandeur, and fine dome-like clouds dazzled and charmed the eye with alternations and brilliant effects of light and shade. The air on descending felt warm.

We were about midway between Wolverhampton and a town (Walsall) when the balloon slightly collapsed, causing it to descend a little, and the shouting of people was plainly heard, who expected the balloon would descend. At 1:48 p.m. sand was discharged, and a very gradual ascent took place, the direction being along the high road to Birmingham. On looking over the side of the car the shadow of the balloon on the clouds was observed to be surrounded by a kind of corona tinted by prismatic colours, and the rippling of the water on the edges of the canal could be seen very distinctly. We discharged sand several times to enable us to rise. The view continued very grand; a great mass of clouds was observed in the east, and a large town lay on our right. The balloon was again full. At 2:34 p.m. and 20 seconds and at 2:45 p.m. thunder was heard from below, but no cloud could be seen. At 2:54 p.m. my pulsations were 100, 107, and 110 successively in one minute. When at the height of 24,000 feet, at 2:59 p.m., a consultation took place as to the prudence of discharging more ballast, or retaining it so as to ensure a safe descent; ultimately it was decided not to ascend, as some clouds whose thickness we could not tell had to be passed through. At 3:03 p.m. it was difficult to obtain a deposit of dew on the hygrometer, and the working of the aspirator became troublesome. A sound like loud thunder was again heard at 3:13 p.m.; at 3:25 p.m. I began to feel unwell. About 3:26 p.m. a most remarkable view presented itself: the sky was of a fine deep blue, dotted with cirri. The earth and its fields, where visible, appeared very beautiful indeed—here, hidden by vast cumuli and plains or seas of cumulo-stratus, causing the country beneath to be shaded for many hundreds of square miles; there, without a cloud to obscure the sun's rays. Again, in other places there were detached cumuli, whose surfaces appeared connected by vast plains of hillocky clouds, and in the interstices the earth was visible, but partly obscured by blue haze or mist. In another place brightly shining cumuli were observed, and seas of detached clouds which cannot be described. Due north, a beautiful cloud, the same we passed through on leaving Wolverhampton, and which had followed us on our way, still reigned in splendour, and might from its grandeur have been called the monarch of clouds. On looking over the top of the car the horizon appeared to be on a level with the eye; the image of the balloon and car, in descending, was very distinctly visible on the clouds. We entered clouds at 3:45 p.m. and lost sight of the sun, but broke through at 3:50 p.m. and saw the earth. Preparations were made for the descent, which, after we had passed through some mist, took place at Solihull, about seven miles from Birmingham.

September 5, 1862

This ascent had been delayed owing to the unfavourable state of the weather. We left the earth at 1:03 p.m.; the temperature of the air was 59°, and that of the dew point 50°. The air at first was misty; at the height of 5,000 feet the temperature was 41°, dew point 37°.9. At 1:13 p.m. we entered a dense cloud of about 1,100 feet in thickness, where the temperature fell to 36.5°, the dew point being the same, thus indicating that the air here was saturated with moisture. At this elevation the report of a gun was heard. Momentarily the clouds became lighter, and on emerging from them at 1:17 p.m. a flood of strong sunlight burst upon us with a beautiful blue sky without a cloud, and beneath us lay a magnificent sea of clouds, its surface varied with endless hills, hillocks, and mountain chains, and with many snow-white tufts rising from it. I here attempted to take a view with the camera, but we were rising with too great rapidity and revolving too quickly to enable me to succeed. The brightness of the clouds, however, was so great that I should have needed but a momentary exposure, Dr. Hill Norris having kindly furnished me with extremely sensitive dry plates for the purpose. We reached the height of two miles at 1:22 p.m., where the sky was of a darker blue, and from whence the earth was visible in occasional patches beneath the clouds. The temperature had fallen to the freezing point, and the dew point to 26°. The height of three miles was attained at 1:28 p.m., with a temperature of 18°, and dew point 13°; from 1:22 p.m. to 1:30 p.m. the wet-bulb thermometer read incorrectly, the ice not being properly formed on it. At 1:34 p.m. Mr. Coxwell was panting for breath; at 1:38 p.m. the mercury of Daniell's hygrometer fell below the limits of the scale. We reached the elevation of four miles at 1:40 p.m.; the temperature was 8°, the dew point minus 15°, or 47° below the freezing point of water. Discharging sand, we in ten minutes attained the altitude of five miles, and the temperature had passed below zero and then read minus 2°. At this point no dew was observed on Regnault's hygrometer when cooled down to minus 30°. Up to this time I had taken observations with comfort, and experienced no difficulty in breathing, whilst Mr. Coxwell, in consequence of the exertions he had to make, had breathed with difficulty for some time. Having discharged sand, we ascended still higher; the aspirator became troublesome to work; and I also found a difficulty in seeing clearly. At 1:51 p.m. the barometer read 10.8 inches. About 1:52 p.m. or later, I read the dry-bulb thermometer as minus 5°; after this I could not see the column of mercury in the wet-bulb thermometer, nor the hands of the watch, nor the fine divisions on any instrument. I asked Mr. Coxwell to help me to read the instruments. In consequence, however, of the rotatory motion of the balloon, which had continued without ceasing since leaving the earth, the valve line had become entangled, and he had to leave the car and mount into the ring to readjust it. I then looked at the barometer, and found its reading to be 9.5 inches, still decreasing fast, implying a height exceeding 29,000 feet. Shortly after I laid my arm upon the table, possessed of its full vigour, but on being desirous of using it I found it powerless—it must have lost its power momentarily; trying to move the other arm, I found it powerless also. Then I tried to shake myself, and succeeded, but I seemed to have no limbs. In looking at the barometer my head fell over my left shoulder; I struggled and shook my body again, but could not move my arms. Getting my head upright for an instant only, it fell on my right shoulder; then I fell backwards, my back resting against the side of the car and my head on its edge. In this position my eyes were directed to Mr. Coxwell in the ring. When I shook my body I seemed to have full power over the muscles of the back, and considerably so over those of the neck, but none over either my arms or my legs. As in the case of the arms, so all muscular power was lost in an instant from my back and neck. I dimly saw Mr. Coxwell, and endeavoured to speak, but could not. In an instant intense darkness overcame me, so that the optic nerve lost power suddenly, but I was still conscious, with as active a brain as at the present moment whilst writing this. I thought I had been seized with asphyxia, and believed I should experience nothing more, as death would come unless we speedily descended: other thoughts were entering my mind, when I suddenly became unconscious as on going to sleep. I cannot tell anything of the sense of hearing, as no sound reaches the ear to break the perfect stillness and silence of the regions between six and seven miles above the earth. My last observation was made at 1:54 p.m. above 29,000 feet. I suppose two or three minutes to have elapsed between my eyes becoming insensible to seeing fine divisions and 1:54 p.m., and then two or three minutes more to have passed till I was insensible, which I think, therefore, took place about 1:56 p.m. or 1:57 p.m.

Whilst powerless I heard the words "temperature" and "observation," and I knew Mr. Coxwell was in the car, speaking to and endeavouring to rouse me—therefore consciousness and hearing had returned. I then heard him speak more emphatically, but could not see, speak, or move. I heard him again say, "Do try; now do." Then the instruments became dimly visible, then Mr. Coxwell, and very shortly I saw clearly. Next I arose in my seat and looked around as though waking from sleep, though not refreshed, and said to Mr. Coxwell, "I have been insensible." He said, "You have, and I too, very nearly." I then drew up my legs, which had been extended, and took a pencil in my hand to begin observations. Mr. Coxwell told me that he had lost the use of his hands, which were black, and I poured brandy over them.

I resumed my observations at 2:07 p.m., recording the barometer reading at 11.53 inches, and temperature minus 2°. It is probable that three or four minutes passed from the time of my hearing the words "temperature" and "observation," till I began to observe; if so, returning consciousness came at 2:04 p.m., and this gives seven minutes for total insensibility. I found the water in the vessel supplying the wet-bulb thermometer one solid mass of ice, though I had, by frequent disturbance, kept it from freezing. It did not all melt until we had been on the ground some time. Mr. Coxwell told me that while in the ring he felt it piercingly cold, that hoar frost was all round the neck of the balloon, and that on attempting to leave the ring he found his hands frozen. He had, therefore, to place his arms on the ring, and drop down. When he saw me he thought for a moment that I had lain back to rest myself, and he spoke to me without eliciting a reply; he then noticed that my legs projected and my arms hung down by my side, and saw that my countenance was serene and placid, without the earnestness and anxiety he had observed before going into the ring: then it struck him that I was insensible. He wished to approach me, but could not; and when he felt insensibility coming over him too, he became anxious to open the valve. But in consequence of having lost the use of his hands he could not do this; ultimately he succeeded, by seizing the cord with his teeth, and dipping his head two or three times, until the balloon took a decided turn downward.

No inconvenience followed my insensibility; and when we dropped it was in a country where no conveyance of any kind could be obtained, so I had to walk between seven and eight miles.

During the descent, which was at first very rapid, the wind was easterly. To check the rapidity of the descent, sand was thrown out at 2:30 p.m. The wet bulb seemed to be free from ice at this time, but I held the bulb between my thumb and finger, for the purpose of melting any ice remaining on it or the connecting thread. The readings after this appeared correct. The final descent took place in the centre of a large grass field belonging to Mr. Kersall, at Cold Weston, seven miles and a half from Ludlow.

I have already said that my last observation was made at a height of 29,000 feet; at this time (1:54 p.m.) we were ascending at the rate of 1,000 feet per minute; and when I resumed observations we were descending at the rate of 2,000 feet per minute. These two positions must be connected, taking into account the interval of time between, viz. thirteen minutes. And on these considerations, the balloon must have attained the altitude of 36,000 or 37,000 feet. Again, a very delicate minimum thermometer read minus 11.9°, and this would give a height of 37,000 feet. Mr. Coxwell, on coming from the ring, noticed that the centre of the aneroid barometer, its blue hand, and a rope attached to the car, were all in the same straight line, and this gave a reading of 7 inches, and leads to the same result. Therefore, these independent means all lead to about the same elevation, viz. fully seven miles.

In this ascent six pigeons were taken up. One was thrown out at the height of three miles, when it extended its wings and dropped like a piece of paper; the second, at four miles, flew vigorously round and round, apparently taking a dip each time; a third was thrown out between four and five miles, and it fell downwards as a stone. A fourth was thrown out at four miles on descending; it flew in a circle, and shortly alighted on the top of the balloon. The two remaining pigeons were brought down to the ground. One was found to be dead; and the other, a carrier, was still living, but would not leave the hand when I attempted to throw it off, till, after a quarter of an hour, it began to peck at a piece of ribbon with which its neck was encircled; it was then jerked off the finger, and shortly afterwards flew with some vigour towards Wolverhampton. One of the pigeons returned to Wolverhampton on Sunday the 7th, and it was the only one I ever heard of.

In this ascent, on passing out of the clouds there was an increase of 9°, and then there was no interruption in the decrease of temperature till the height of 15,000 feet was reached, when a warm current of air was entered, which continued to 24,000 feet, after which the regular decrease of temperature continued to the highest point reached. On descending, the same current was again met with, between 22,000 and 23,000 feet. A similar interruption, but to a greater amount, was experienced till the balloon had descended to about the same height in which it was reached on ascending; after this no further break occurred in the regular increase of temperature, the sky being clear till the descent was completed. From the general agreement of the results as observed by Regnault's hygrometer, and those of the dew point as found by the dry and wet bulb thermometers, there can be no doubt that the temperature of the dew point, at heights exceeding 30,000 feet, must have been as low as minus 50° below the zero of Fahrenheit's scale, or 82° below the freezing point of water, implying that the air was very dry.

Chapter IV

Ascents from the Crystal Palace

April 18, 1863

In this ascent the balloon was partially filled during the evening of April 17, with the view of starting early the following morning. The atmosphere was at this time thick and misty; the wind on the earth was N.E., but pilot balloons on attaining a moderate elevation fell into a north current. The wind was moving at an estimated velocity of forty miles an hour, and the ascent was delayed hour after hour, in the hopes that the upper current would change to N.E. At 1 p.m., when the sky was nearly covered with clouds, and there were occasionally gleams of sunshine, the ascent was decided upon, although it was evident it could not be one of long duration, unless the wind changed its direction, or we resolved to cross the Channel. Whilst discussing this, the rope, our only connecting link with the earth, broke, and at 1:17 p.m. we started very unceremoniously, the balloon taking a great lurch; I was thrown among my instruments, and unfortunately both Daniell's and Regnault's hygrometers were broken. Within three minutes we were more than 3,000 feet high. At 4,000 feet, cumulus clouds were on our level, and a thick mist rested everywhere on the earth. At 1:26 p.m. we were 7,000 feet high, in a thick mist which almost amounted to a fog. The temperature of the air continued at 32° nearly, whilst that of the dew point increased several degrees. On passing out of the cloud these two temperatures very suddenly separated, the latter decreasing rapidly; the sky was of a deep blue, without a cloud on its surface. At 1:30 p.m. we were 10,000 feet high; directly under us was a sea of clouds. The towers of the Crystal Palace were visible, and by them we found we were moving south.

The temperature before starting was 61°; it decreased to 32° on reaching the cloud, and continued at this value whilst in it; then suddenly fell to 23.25° on leaving the cloud, and was either less or the same at every successive reading till we reached the height of 20,000 feet, where the lowest temperature was noticed. In passing above four miles the temperature increased to 14.25°, and then declined to 12.25° at the highest point, viz. 24,000 feet, at one hour and thirteen minutes after starting. When we were just four miles high, on descending, we began to reflect that possibly we might have been moving more quickly than we expected, and it was necessary to descend till we could see the earth below. The valve was opened rather freely at 2:34 p.m., and we fell a mile in three minutes. We descended quickly, but less rapidly, through the next mile, and reached the clouds at 12,000 feet from the earth, at 2:42 p.m. On breaking through them at 2:44 p.m., still 10,000 feet from the earth, I was busy with my instruments, when I heard Mr. Coxwell exclaim, "What's that ?" He had caught sight of Beachy Head. *I looked over the car, and the sea seemed to be under us.* Mr. Coxwell again exclaimed, "There's not a moment to spare; we must save the land at all risks. Leave the instruments." Mr. Coxwell almost hung to the valve line, and told me to do the same, and not to mind its cutting my hand. It was a bold decision, opening the valve in this way, and it was boldly carried out.

When a mile high, the earth seemed to be coming up to us. There were two rents in the balloon, cut by the valve line; these we could not heed. Up, up, the earth appeared to come, the fields momentarily enlarging; and we struck the earth at 2:48 p.m. at Newhaven, very near the sea—of course with a great crash, but the balloon by the very free use of the valve line had been crippled and never rose again, or even dragged us from the spot on which we fell. Nearly all the instruments were broken, and to my great regret three very delicate and beautiful thermometers, specially sent to me by M. A. d'Abbadie for these observations, were all broken. I was fortunate, however, in seizing and pocketing the aneroid barometer which had been up with me in every high ascent. It was this instrument that Mr. Coxwell read when we were seven miles high, and I at the time in a state of insensibility.

The ascent was gradual from 16,000 feet to the highest point, and there was sufficient time for the instruments to attain the true temperature. We were above four miles for half an hour, not passing above 24,000 feet. On passing below four miles it would seem that the drop to three miles was nearly a straight line, and the next mile, though occupying a little more time, was passed quickly. The position of the clouds was fortunately very high, as well as the very rapid descent of two miles in four minutes. The whole time of descending the four miles and a quarter was about a quarter of an hour only.

July 11, 1863

The ascent from the Crystal Palace, July 11, was intended to have been one of extreme height, and the promise of success in this respect was held out until near the time of starting, as pilot balloons had passed nearly due east, and indicated that our course would have been towards Devonshire; but so doubtful is the course a balloon will take, that no certainty can be felt till the balloon has actually left. However, on this occasion pilot balloons, though at first moving towards the east, soon met with a north wind and went south.

Under these circumstances, the attempt to ascend five miles was abandoned, and we resolved to ascertain, as far as possible, the thickness of the stratum influenced by the east wind, and if possible to profit by the knowledge and have as long a journey as we could.

At the time of leaving, 4:55 p.m., the sky was nearly covered with cirrus and cirrostratus clouds, and the wind was blowing due east.

In about four minutes, and when at the height of 2,400 feet, the balloon suddenly changed from moving towards the west, to moving due south. At 5:08 p.m. we were over Croydon, at the height of 4,600 feet, in mist, but could see the Green Man Hotel, Blackheath; we then descended, passing downward through a thick atmosphere, till at 5:32 p.m. we were 2,200 feet high over Epsom Downs, and again within the influence of the east wind. We then turned to ascend, and at 5:52 p.m. were 3,000 feet above Reigate, and we here could see Shooter's Hill and the Crystal Palace, by the two towers of which we found we were again within the influence of a north wind.

We then continued to ascend, with the view of ascertaining if we could pass above the stratum which was under the influence of the north wind, at 6:16 p.m.; when at 5,400 feet, the wind shifted to N.N.W., and the atmosphere became very thick and misty, the sun's place being just visible. At 6:28 p.m. we were 6,600 feet high, and the sun was wholly obscured; we descended somewhat, but did not get below the mist. At 6:40 p.m. we were 6,200 feet high, and directly over Horsham; and here I essayed to take a photograph, but from the mist by which we were surrounded, and the dark earth below—not lighted up at all—I did not succeed. We then ascended to 6,600 feet again to repeat the observations I had made, and found that the temperature at this elevation in the half-hour had declined 2° or 3°.

At this time, 6:56 p.m., cirri and cirrostratus were very much higher than ourselves, and we saw the coast near Brighton. A consultation had been held whilst at this height, with the view of crossing over to France, but our progress being so slow, the circumstances did not promise success, so we came down with the view of again falling into the east wind, supposing it still to be prevalent. We met the north wind again at about 5,000 feet, and the east wind at exactly the same height, viz. 2,400 feet, at which we lost it on ascending. We descended to within 1,000 feet of the earth, and were near Worthing, at about five miles from the coast. We then ascended to 2,700 feet, and found ourselves moving towards the coast, and therefore within the influence of a north wind; evidently, therefore, if we wished to continue our journey, we must keep below 2,400 feet, otherwise we should be blown out to sea. When again at the height of 2,400 feet, we turned to move parallel to the coast, being at this time over Arundel. Sheep in the fields were evidently very frightened, and they huddled together. We now descended to 800 feet, and thus journeyed at heights varying from 800 to 1,600 feet, villagers frequently shouting to us to come down, and now and then answering our questions as to the locality we were in. The cheering cry of children was frequently heard above all other sounds. Geese cackled, and, frightened, scuttled off to their farms; pheasants crowed as they were going to roost, and as we approached the end of our journey packs of dogs barked in the wildest state of excitement at the balloon. Journeying in this way was most delightful; all motion seemed transferred to the landscape itself, which appeared when looking one way to be rising and coming toward us, and when looking the other as receding from us. It was charmingly varied with parks, mansions, and white roads, and in fact all particulars to make up a rural scene of character extremely beautiful.

The temperature of the air was 75° on the ground, decreased to 63° at 2,600 feet, differed but little from 62° between 2,800 and 3,400, and then declined gradually to 55° at 5,000 feet; at heights exceeding 5,600 feet the temperature differed but little from 53°.

The direction of the wind on the ground was east, at 2,600 feet it was north, and at heights exceeding 5,400 feet it was N.N.W.

The humidity of the air on passing from the east wind at 2,400 feet, to the north wind, increased greatly, and continued to increase till nearly 5,400 feet, when the direction of the wind changed to N.N.W., and at heights greater than this there were no clouds, but the air was very misty.

When we were at the height of 2,600 feet, flat-bottomed cumulus clouds were at our level. The clouds were entirely within the influence of the north wind, their undersides were in contact with the east wind, with a much drier air, which at once dissipated all vapour in contact with it, and thus prevented the appearance of flat-bottomed clouds.

My friend Mr. Nasmyth, in a letter to me, says: "The flatness of the undersides of the clouds during settled weather appears to me to rest on the upper surface of a stratum of air which appears to terminate at the line of flat bottom of the cloud." And these are the exact circumstances in which on this occasion I saw them.

Chapter V

Ascent from Wolverton

June 26, 1863

In the ascent from Wolverton on June 26, the Directors of the North Western Railway Company provided the gas, and gave every facility to members of the Committee of the British Association and their friends to be present. The gasometers at Wolverton are too small to hold gas enough to fill the balloon: it was therefore partly inflated the night before, and remained out all night without being influenced by the slightest wind. The morning of the ascent was also calm; the sky was of a deep blue, implying the presence of but little vapour. The atmosphere was bright and clear, and all the circumstances were of the most promising kind. The time of ascent was fixed to take place some little time after the express train from London should arrive, or at a little after noon; and the filling was somewhat delayed, the extraordinary fineness of the morning promising its completion in a short time. Between eleven and twelve all these favourable circumstances changed; the sky became covered with clouds, and some of them of a stormy character. The wind arose and blew strongly; the balloon lurched a great deal. Great difficulty was experienced in passing the gas into the balloon, and sufficient could not be passed in by one o'clock. The wind was momentarily increasing, and it became very desirable to be away. The greatest difficulty was experienced in fixing the instruments, and some were in great danger of being broken by the violent swaying of the balloon and the incessant striking of the car upon the ground, notwithstanding the exertions of fifty men to hold it fast. At the time of leaving, the spring catch was jammed so tight by the pressure of the wind that it would not act, and we were let free by the simultaneous yielding of the men, and had to part instantly with ballast to avoid striking adjacent buildings.

It was three minutes after one when we left the earth, with a strong W.S.W. wind: the temperature 65°. In four minutes we were 4,000 feet high, and entered a cloud with a temperature of 50°, experiencing a most painful feeling of cold. As on all previous occasions, we expected soon to break through the clouds into a flood of strong sunlight, with a beautiful blue sky, without a cloud above us, and with seas of rocky clouds below. But, on the contrary, when we emerged, it was dark and dull. Above us there were clouds. At 9,000 feet high we heard the sighing, or rather moaning of the wind as preceding a storm: it was the first time that I had heard such a sound in the air. We satisfied ourselves that it was in no way attributable to any movement of the cordage about the balloon, but that it was owing to conflicting currents of air beneath. At this time we saw the sun very faintly, and momentarily expected its brilliancy to increase, but instead of this, although we now were two miles high, we entered a fog, and entirely lost the sight of it. Shortly afterwards fine rain fell upon us. Then we entered a dry fog, and at 12,000 feet passed out of it; saw the sun again faintly for a short time, and then entered a wetting fog. At 15,000 feet we were still in fog, but it was not so wetting. At 16,000 feet we entered a dry fog; at 17,000 feet saw faint gleams of the sun, and at the same height we heard a train. We were now about three miles high. As we looked around there were clouds below us, others on our level at a distance, and yet more above. We looked with astonishment at each other, and said that as we were rising steadily we must surely soon pass through them.

At 17,500 feet we were again enveloped in fog, which became wetting at 18,500 feet. We left this cloud below at 19,600 feet. At 20,000 feet the sun was just visible. We were now approaching four miles high; clouds, dense clouds, were still above us; for a space of 2,000 to 3,000 feet we met with no fog, but on passing above four miles high our attention was attracted to a dark mass of cloud, and then to another on our level. Both these clouds had fringed edges, and were unmistakeably nimbi. Without the slightest doubt they were both rain clouds. Whilst looking at them we again lost sight of everything, being enveloped in fog whilst passing upwards through 1,000 feet. At 22,000 feet we emerged again, and were above clouds on passing above 23,000 feet. At six minutes to two we heard a railway train: the temperature here was 18°. I still wished to ascend to find the limits of this vapour, but Mr. Coxwell knew better, and I was met with a negative: "Too short of sand. I cannot go higher; we must not even stop here." I was therefore most reluctantly compelled to abandon the wish, and looked searchingly around. At this highest point, in close proximity to us, were rain clouds;

below us, dense fog. I was again reminded that we must not stop here. With a hasty glance everywhere, above, below, all around, I saw the sky nearly covered with dark clouds of a stratus character, with cirri still higher, and small spaces of faint blue sky between them; the blue was not the blue of four or five miles high, as I had always before seen it, but a faint blue, as seen from the earth when the air is charged with moisture.

Hastily glancing over the whole scene, there was no extensive, fine, or picturesque views, as in such situations I had always before seen. The visible area was limited; the atmosphere was murky; the clouds were confused, and the aspect everywhere dull.

I cannot avoid expressing the surprise I have felt at the extraordinary power which a situation like this calls forth, when it is felt that a few moments only can be devoted to note down all appearances and all circumstances at these extreme positions; and if not so rapidly gleaned, they are lost forever. In such situations every appearance of the most trivial kind is noticed; the eye seems to become keener, the brain more active, and every sense increased in power to meet the necessities of the case; and afterwards, when time has elapsed, it is wonderful how distinctly, at any moment, scenes so witnessed can be recalled and made to reappear mentally in all their details, so vividly, that had I the power of the painter I could reproduce them visibly to the eye upon the canvas.

We then began our downward journey, wondering whether we should meet the same phenomena. Soon we were enveloped in fog, but passed below it when at 22,000 feet, and then we saw the sun faintly. At 20,000 feet we were in a wetting fog, and passed beneath it at 19,500 feet, experiencing great chilliness; fog was then above and below. I now wished to ascend into the fog again, to check the accuracy of my readings as to its temperature, and the reality of the chill we had felt. This we did: the temperature rose to its previous reading, and fell again on descending.

For the next 1,000 feet we passed down through a thick atmosphere, but not in cloud or fog. At the height of 18,000 feet we were again in fog. At three miles high we were still in fog, and on passing just below three miles, rain fell pattering on the balloon. This was one mile higher than we experienced rain on the ascent, but it was much heavier. On passing below 14,000 feet, and for a space of nearly 5,000 feet, we passed through a beautiful snowy scene. There were no flakes in the air—the snow was entirely composed of spiculae of ice, of cross spiculae at angles of 60°, and an innumerable number of snow crystals, small in size, but distinct and of well-known forms, easily recognizable as they fell and remained on the coat. This unexpected meeting with snow on a summer afternoon was all that was needed on this occasion to complete the experience of the characteristics of extreme heat of summer with the cold of winter within the range of a few hours. On passing below the snow, which we did when about 10,000 feet from the earth, we entered a murky atmosphere which continued till we reached the ground; indeed, so thick and misty was the lower atmosphere, that although we passed nearly over Ely Cathedral and not far from it, we were unable to see it. When 5,000 feet high, we were without sand and simply became a falling body, the rapidity of the fall being checked by throwing the lower part of the balloon into the shape of a parachute. The place of descent was in a field on the borders of the counties of Cambridge and Norfolk, twenty miles from the mouth of the Wash, and eight miles from Ely.

This Wolverton ascent must rank among the most extraordinary of my series, giving scientific data of high interest and results most unexpected.

Ascent from the Crystal Palace

July 21, 1863

The weather on this day was bad, the sky overcast and rainy. Although in every respect a thoroughly bad day, it was well suited to a particular purpose I had in view, viz.: to investigate, if possible, some points concerning the formation of rain in the clouds themselves; to determine why a much larger amount of rain is collected in a gauge near the surface of the earth than in one placed at an elevation in the same locality, and whether during rain the air is saturated completely; or, if not, to what extent; also to discover the regulating causes of a rainfall, which sometimes occurs in large drops, at others in minute particles.

So long back as the years 1842 and 1843 I made many experiments in order to ascertain why so great a difference in volume was found to exist in the water collected at lower stations as compared with that collected at higher.

The experiments which yielded the best results were those in relation to temperature.

I always found that when the rain was warm, with respect to

the temperature of the air at the time, no difference existed in the quantities of rain collected at different heights; but when the temperature of the rain was lower than the temperature of the air, a considerable difference existed.

From this circumstance it would appear probable that the difference in the quantities of rain collected at different heights is owing (at least in part) to the great condensation of the vapour in the lower atmosphere, through being in contact with the relatively cold rain.

In this ascent I desired to confirm or otherwise Mr. Green's deductions.

This gentleman believing that whenever a fall of rain happens from an overcast sky there will invariably be found to exist another stratum of cloud at a certain elevation above the first, I determined, if I found it so, to measure the space between them and the thickness of the upper stratum, and to ascertain whether the sun was shining on its upper surface.

We left the earth at 4:52 p.m., and in ten seconds had ascended into the mist; in twenty seconds, to a level with the clouds, but not through them. At the height of 1,200 feet we passed out of this rain.

At the height of 2,800 feet we emerged from clouds, and saw a stratum of darker cloud above; we then descended to 800 feet, over the West India Docks, and saw rain falling heavily upon the earth. None fell upon the balloon; that which we saw, therefore, had its origin within 800 feet of the ground.

We ascended again, and this time passed upwards through fog 1,400 feet in thickness.

At 3,300 feet we were out of cloud, and again saw the dark stratum at a distance above; clouds obscured the earth below.

On descending, at 2,700 feet we entered a dry fog, but it became wetting 100 feet lower down. After passing through 600 feet of it, the clouds became more and more wetting, and below were intensely black.

At 5:28 p.m. we were about 700 feet high, or about 500 feet above Epping Forest, and heard the noise of the rain pattering upon the trees.

Again we ascended to 2,000 feet; then through squalls of rain and wind descended to 200 feet, the raindrops being as large as a fourpenny piece, the same as when we left the earth.

On reaching the earth, we found that rain had been falling heavily all the time we were in the air.

Thus this journey gave more information about rain than we ever before had gained, and which could be obtained by means of the balloon alone.

Chapter VI

Ascent from Windsor

May 29, 1866

No ascent had been made in May, and I was anxious to make one in this month. Mr. [Henry Emerson] Westcar, of the Royal Horse Guards, then stationed at Windsor, kindly offered the use of his balloon, and arrangements were made for ascents at different times in May, but, as is usual, some fruitless attempts were made.

On May 29 the balloon was filled early in the afternoon, and we left at 6:14 p.m., about an hour and three-quarters before sunset, in the hope of being able to remain in the air for as long a time after sunset.

The temperature of the air at this time was 58°, and 58.5° at Greenwich Observatory. It at once declined on leaving the earth to 55° at 1,200 feet, and to 43° between the heights of 3,600 to 4,600 feet, then further declining to 29.5° at the height of 6,200 feet, at 7:17 p.m. On descending, the temperature increased, but not uniformly, to 54° at 8:09 p.m. at 380 feet above the sea, when, however, we were nearly touching the tops of the trees, there being about 3° of less temperature when at the same height above the sea on rising. Our object was to be as near the earth as possible at the time of sunset, and, afterwards, to discharge sand so quickly as to see sunrise again in the west. We did not succeed. At the time of sunset we were about 600 feet high, but had just passed over a hill, and on passing the ridge the balloon had been sucked down, so that it was only by a free discharge of sand that Mr. Westcar prevented the balloon coming to the ground. We then again started upon a second ascent, to be as like the one we had just completed as we could make it. At 8:09 p.m. the temperature was 54°. Again the temperature declined, but somewhat less rapidly than before. On again reaching one mile the temperature had declined to 39°, and on reaching the height of 6,200 feet (the same elevation as we were three-quarters of an hour before sunset), the sun having set nearly twenty minutes, the temperature was 35°, or about 6° warmer than when at the same elevation something more than one hour before. On descending, the temperature changed very little, being 35° to 36° for a thousand feet downwards. It increased to 37° at 4,500, to 47° at 1,500, and to 54° at 900 feet; but here the increase was checked, and at 600 feet the temperature was 52.75°; on ascending a little, again the temperature increased, it decreased on descending, and was 50.25° on the ground at a spot 300 feet above the sea, at half-past eight o'clock. At Greenwich at this time the temperature of the air was 52°.

At the time of leaving the earth at 6:14 p.m. the air at Greenwich had but 3 grains of moisture in a cubic foot. At Windsor, near the Thames, there were 4.5 grains; the air was damp: on ascending the air at first became drier, but at the height of one mile was saturated, and was very nearly saturated at the same height after sunset.

Thus this double ascent enables us to compare the temperatures of the same elevations, just before and just after sunset on the same day, and to estimate the amount of heat radiated from the earth at about the time of sunset.

At heights exceeding 2,000 feet the direction of the wind was N. by W.; at the height of one mile the air was nearly calm; and at heights less than 2,000 feet it was N. by E., and these currents were met with always at those elevations.

At all times during the ascent, whenever the sun shone upon a transparent bulb, or a dull blackened bulb thermometer, the reading was a very little in excess of the reading of a shaded bulb, and was frequently the same even when the sun's heat felt sensibly warm.

The path of the balloon from Windsor was over Windsor Great Park; nearly over Woking at 7:43 p.m.; a little west of Guildford, where, approaching the coast, at half-past nine, we calculated that the sea must be near, and we descended at a place five miles south of Pulborough.

My attention was almost wholly occupied with the observations; Mr. Westcar's chiefly with the management of the balloon: he frequently, however, read the several instruments, particularly those whose bulbs were exposed to the sun's rays.

The safety lamp was burning all the time, thus enabling the instruments to be read after dark.

I till recently believed that this was the first ascent for scientific purposes, since that of Biot and Gay-Lussac in 1804, in which the management of the balloon was undertaken by the experimentalists themselves. But I find I am in error in this respect. My friend l'Abbé [Françios Napoléon Marie] Moigno tells me that Messrs Bixio and Barral, in the year 1850, took the entire management of the balloon in their own hands.

On descending, nearly one hour and a half after sunset, there was no one near to assist us to empty and pack the balloon. This we had to do ourselves, and we were preparing to pass the night in the car of the balloon, when towards midnight a shepherd came by, and we passed the night in his cottage at the distance of half a mile, leaving the balloon, etc., in the fields till the morning.

The temperature of the air declined from 58° on the ground to 52° at 2,000 feet, and somewhat more rapidly to 46.7° at 3,000 feet; it increased to 48.7°, or by 2° in the next 400 feet, and then gradually declined to 29.8° at the height of 6,200 feet. On descending, the temperature increased gradually to 48.3° at 1,000 feet, and then much more rapidly to 53.6° at the height of 500 feet: this rapid increase was remarkable. On turning to ascend, the sun having set the temperature declined pretty equally to the height of 4,000 feet, and at greater heights with somewhat less regularity, to 34° at 6,000 feet, when the temperature increased to 35.3° at the height of 6,400 feet: this increase was very remarkable. On descending again, the temperature increased with moderate regularity to 48.7° at the height of 1,300 feet, and then with much greater rapidity to 53.8° at the height of 600 feet, when the increase was arrested, and the temperature at lower elevations rapidly declined to 50.1° on reaching the earth. This decline of temperature from 600 feet is remarkable. By comparing the readings at the same heights before and after sunset, it will be seen that at the height of 6,000 feet the temperature was from 5° to 6° warmer after sunset than it was before sunset, and that the temperatures on the ground and at 1,000 feet high were nearly the same, whilst at intermediate heights they were much higher.

The degree of humidity of the air increased from the ground to the height of 500 feet; from this height to 1,200 feet the air was somewhat less humid, and still less so at heights exceeding 1,200 feet. At the height of 3,400 feet the degree of humidity was 57 only; the air was again wet at 4,800 feet, and somewhat less so at heights exceeding 5,000 feet. On descending, the humidity of the air was more uniform down to the height of 3,400 feet, and below this the air was less humid than at the same elevations on the ascent, particularly at low elevations. On descending below 400 feet, I packed up the instruments, for fear of the balloon striking the ground; at this time the sun was setting. On ascending again, after sunset, the air was more and more humid, and most so at 6,300 feet; the same result we found in the descent, to the height of 600 feet, where the degree of humidity was 61; and it increased to 68 on the ground.

Chapter VII

Over London by day

March 31, 1863

The day was favourable; the wind was from the east, in gentle motion; the sky was blue and almost cloudless. The earth was left at 4:16 p.m., and we passed upwards with very nearly an even motion to the height of 19,000 feet, continued at about this level for some little time, and then gradually ascended to 24,000 feet, which we reached at 5:28 p.m., or in one hour and twelve minutes after starting. We then let out gas, and never have I seen the opening of the valve exercise such an effect, for though it seemed to be but momentary, we fell in consequence a mile and a quarter in four minutes. Happily we had enough sand to contend with this difficulty, and checked the descent by parting with it, and for half an hour we kept nearly upon the same level, between 15,000 and 16,000 feet high. After this we gradually and almost continuously fell, and reached the earth at 6:26 p.m., effecting the descent in fifty-eight minutes from the place where the balloon was at its secondary station.

The temperature of the air on the earth on leaving was 50°. At 4:25 p.m., at the height of one mile, it was 33.5°; the second mile was reached at 4:35 p.m., with a temperature of 26°; the third mile at 4:44 p.m., when the temperature was 14°; and at 3.75 miles high the temperature was 8°. A warm current of air was here met with, and the temperature rose to 12° at 4:58 p.m.; at 5:02 p.m. the warm current was passed, and when 4.5 miles high the temperature was just zero of Fahrenheit's scale.

In descending, the temperature increased to 11°, at about three miles high; then a cold current was met with, and it fell to 7°. This was soon passed, and the temperature increased to 18.5° at two miles high, to 25.5° at one mile, and to 42° on the ground, showing a decrease of 8° of temperature during the two hours and ten minutes between the two observations. On comparing the readings of thermometers at the same height during the ascents with those during the descents, all the latter were lower, showing that the whole mass of air was of lower temperature than that in immediate contact with the earth, but to a smaller amount. The air was dry before leaving; it became very dry at heights exceeding two miles, and at heights exceeding four miles the temperature of the metallic cup of Regnault's hygrometer was lowered to nearly minus 40°, and no dew was deposited on its surface. The temperature of each layer of air was different, according to its direction of motion, and there were several currents met with. Within two miles of the earth the wind was east; between two and three miles high it was directly opposite, viz. west. About three miles it was north-east; higher still it changed to the opposite—south-west; and from about four miles to the highest point reached, it was west.

We left the Crystal Palace, therefore, with an east wind, and at about 4:48 p.m. the Palace appeared directly under us.

When one mile high the deep sound of London, like the

roar of the sea, was heard distinctly; its murmuring noise was heard at greater elevations. At the height of three and four miles the view was indeed wonderful: the plan-like appearance of London and its suburbs; the map-like appearance of the country generally; and the winding Thames, leading the eye to the white cliffs at Margate and on to Dover, were sharply defined. Brighton was seen, and the sea beyond, and all the coast line up to Yarmouth. The north was obscured by clouds. Looking underneath, and to the south, there were many detached cumuli clouds, and in some places a solitary cloud; ail apparently resting on the earth. Towards Windsor the Thames looked like burnished gold, and the surrounding water like bright silver. Railway trains were like creeping things, caterpillar-like, and the steam like a narrow line of serpentine mist. All the docks were mapped out, and every object of moderate size was clearly seen with the naked eye. Taking a grand view of the whole visible area beneath, I was struck with its great regularity: all was dwarfed to one plane; it seemed too flat, too even, apparently artificial. The effect of the river scenery in this respect was remarkable; the ships, visible even beyond the Medway, looked like toys.

At the height of three miles and a half Mr. Coxwell said my face was of a glowing purple, and higher still both our faces were blue. At heights exceeding three miles, our feet and the tips of our fingers were very cold. The sky was of a deep Prussian blue. When three miles high, on descending, Mr. Coxwell, forgetful of the fact that the grapnel had been exposed to a temperature of zero, incautiously took hold of it with his naked hand, and cried out, as in pain, that he was scalded, and called on me to assist in dropping it. The sensation was exactly that of scalding. The blackness creeping over the land at sunset, whilst the sun was still shining on us, was remarkable. We reached the ground at 6:03 p.m., near Barking, in Essex.

Over London by night

Ascent from Woolwich Arsenal, October 2, 1865

When the sun had set for nearly three-quarters of an hour, and night had fairly set in, the moon shining brightly, and the sky free from cloud, the balloon left Woolwich Arsenal at 6:20 p.m., the temperature at the time being 56°. Within three or four minutes a height of 900 feet was reached, and till this time I had failed in directing the light of the Davy lamp properly. When I succeeded, the temperature was 57° and increasing; on reaching 1,200 feet high it had increased to 58.9°. We then descended to 900 feet, and the temperature decreased to 57.8°; on beginning to ascend again the temperature increased to 59.6° at 1,900 feet high, being 3.5° warmer than when the earth was left. On descending again the temperature decreased to 57.5° at the height of 600 feet, and in the several subsequent ascents and descents the temperature increased with elevation, and decreased on approaching the earth. On every occasion the highest temperature was met with at the highest point. This result was remarkable indeed. The different degrees of the humidity of the air met with in this ascent are no less remarkable. Considering saturated air as represented by 100, at the commencement of the ascent in the balloon it was 95; at Greenwich Observatory it was 84; towards the end of the ascent in the balloon it was 85, and at Greenwich was 97. The state of things was therefore reversed, and would indicate that the water in the air had fallen. Its amount at the beginning of the ascent was 5.25 grains in a cubic foot of air, and at the same elevation was 4.5 grains in the same mass of air at the end of the ascent.

The readings of the instruments were taken very slowly, owing to the difficulty experienced in directing the light properly. I failed in all magnetic experiments, and indeed in nearly all but those relating to temperature and humidity. Two self-registering minimum thermometers were tied down, one with its bulb resting on cotton wool, fully exposed to the sky, and the other with its bulb projecting beyond the supporting frame; their indexes were at the end of their columns of spirit on starting, or at 56°. At every examination of each of these instruments a space was found between its index (which remained unmoved) and the end of the column of spirit, indicating a temperature higher than before leaving, and it was closely approximate at all times to the temperature of the air. Consequently, notwithstanding the clearness of the sky, the loss of heat by radiation must have been small. No ozone was shown at the Royal Observatory, but in the balloon paper tests were coloured to 4, on a scale where greatest intensity was considered 10.

At the early part of this ascent I was wholly occupied with the instruments, and when at the height of about 1,000 feet the view which suddenly opened far exceeds description. Almost immediately under, but a little to the south-east, was Woolwich; north was Blackwall; south, Greenwich and Deptford; and west, as far as the eye could reach, was London—the whole forming a starry spectacle of such brilliancy as far to exceed anything I ever saw. When I have been at this elevation in the evening, at a distance from London, it has had the appearance of a vast conflagration, but on this night the air was so clear and free from haze that each and every light was distinct, and they seemed all but touching each other.

The whole of Woolwich, Blackwall, Deptford, and Greenwich could be traced as a perfect model by the line of lights of their streets and squares. In nine minutes we were opposite Brunswick Pier, Blackwall, crossing the Thames; we then passed across the Isle of Dogs, Greenwich Reach, and so up the river Thames. As we advanced towards London, the mass of illumination increased in intensity. At 6:42 p.m. the South Eastern Railway Terminus at London Bridge was directly under us; looking southward at this time we saw the Borough stretching far away, and the many streets shooting from it, particularly Southwark Street, with its graceful curve of lamps. In one minute more we were over Southwark Bridge, 1,300 feet high, passed Blackfriars Bridge at 6:45 p.m., and Charing Cross at 6:47 p.m.

On leaving Charing Cross I looked back over London, the model of which could be seen and traced—its squares by their lights; the river, which looked dark and dull, by the double row of lights on every bridge spanning it. Looking round, two of the illuminated dials of Westminster clock were like two dull moons. Again, looking eastward, the whole lines of the Commercial and Whitechapel roads, with their continuations through Holborn to Oxford Street, were visible, and most brilliant and remarkable. We were at such a distance from the Commercial Road that it appeared like a line of brilliant fire, assuming a more imposing appearance when the line separated into two, and most imposing just under us in Oxford Street. Here the two thickly-studded rows of brilliant lights were seen on either side of the street, with a narrow dark space between, and this dark space was bounded, as it were, on both sides by a bright fringe like frosted silver. At first I could not account for this appearance; but presently, at one point more brilliant than the rest, persons were seen passing, their shadows being thrown on the pavement, and at once it was evident this rich effect was caused by the bright illumination of the shop lights on the pavements.

I feel it impossible to convey any adequate idea of the brilliant effect of London, viewed at an elevation of 1,300 feet, on a clear night, when the air is free from mist.

It seemed to me to realize a wish I have felt when looking through a telescope at portions of the Milky Way, when the field of view appeared covered with gold-dust, to be possessed of the power to see those minute spots of light as brilliant stars; for certainly the intense brilliancy of London this night must have rivalled such a view.

We were over the Marble Arch at 6:51 p.m., about eleven miles in a straight line from Woolwich, which distance had been passed in about half an hour. We therefore were travelling at more than twenty miles an hour. On passing onwards we left the Edgeware Road on our right; and the Great Western Railway on our left, and passed nearly down the Harrow Road. In six or seven minutes we left the suburbs of London, passing over Middlesex in the direction of Uxbridge: there the contrast was great indeed; not a single object could anywhere be seen, not a sound reached the ear; the roar of London was entirely lost. The moon was shining, but seemed to give no light; and the earth could not be seen. After a time the moon seemed to shine with increased brightness; the fields gradually came into view, then the shadow of the balloon on the earth was seen distinctly pointing out our path, which, by reference to the pole star and the moon, became well known to us. After this, occasional masses of lights appeared as we passed over towns and villages. Thus we passed out of Middlesex, over parts of Buckinghamshire and Berkshire, to Highmoor, in Oxfordshire, where we descended on the farm of Mr. Reeves at 8:20 p.m., distant about forty-five miles from Woolwich. The horizontal movement of the air at Greenwich in the same time was registered as sixteen miles.

Unfortunately, Mr. [Alfred] Orton believed we were near the sea, and, notwithstanding my assertions and assurances to the contrary, he suddenly brought the balloon to the ground, and broke nearly all the instruments; the lamp was lost, but an offered reward brought it to me a fortnight afterwards in a very battered condition.

The results of this first night experiment are very valuable; and, so far as one experiment can give, indicate that, on a clear night, the temperature, up to a certain elevation, increases with increase of elevation.

The temperature of the dew point increased on ascending to the height of 900 feet, then decreased, the air becoming drier, or the degree of humidity less; at heights exceeding 1,200 feet the degree of humidity was nearly the same as at heights less than 900 feet. On descending, the temperature of the dew point decreased, and the air was driest at about the height of 1,000 feet; at heights less than 1,000 feet the temperature of the dew point increased, and the degree of humidity increased till the ground was reached.

The temperature of the air was the lowest on the ground, and increased with elevation to the height of 2,000 feet, the highest point attained; and on the descent it decreased with decrease of elevation, and was lowest on reaching the ground.

Chapter VIII

Decrease of temperature with elevation

The few ascents which I have chosen are sufficient to show that the decrease of temperature is very far from constant. It follows, therefore, that we must entirely abandon the theory of a decline of one degree of temperature for every increase of 300 feet of elevation. It is necessary to renounce this ideal regularity upon which we have been dependent in determining the co-efficient of refraction. The differences have been immense; even with a clear sky, the most favourable for establishing a mean, the figures vary very greatly—that is to say, within 100 feet near to the earth we now know there may be a decline of temperature of several degrees during the mid-hours of the day, and that during the mid-hours of the night there may be, and generally is, an increase of several degrees.

The decline of temperature near the earth was found to be different, according to the more or less cloudy state of the sky, being more rapid when the latter was clear than when cloudy; it was, therefore, found necessary to separate the experiments made in one state of the sky from those made in the other. Collecting the results together, the general result of all the mid-day experiments are as follows:—

The change from the ground to 1,000 feet high was 4.5° with a cloudy sky, and 6.2° with a clear sky. At 10,000 feet high it was 2.2° with a cloudy sky, and 2° with a clear sky. At 20,000 feet high the decline of temperature was 1.1° with a cloudy sky, and 1.2°; with a clear sky. At 30,000 feet high the whole decline of temperature was found to be 62°. Within the first 1,000 feet the average space passed through for 1° was 223 feet with a cloudy sky, and 162 feet with a clear sky. At 10,000 feet the space passed through for a like decline was 455 feet for the former, and 417 feet for the latter; and above 20,000 feet high the space with both states of the sky was 1,000 feet nearly for a decline of 1°.

As regards the law thus indicated, it is far more natural and far more consistent than that of a uniform rate of decrease. The results here spoken of have relation to experiments made during the hours of the day; near the earth they do not hold good during the hours of the night, nor are they of universal application during the day, as the following experiences will prove.

In my ascent on January 12, 1864, the temperature of the air before starting was 41.5°; it then decreased very slowly till 1,300 feet was reached, when a warm current was met with, and at 3,000 feet the temperature was 45°, being 3.5° warmer than on the ground, and for the next 3,000 feet the temperature was higher than on the earth. It then gradually fell to 11° at 11,500 feet, and remained at this reading till 12,000 feet was reached.

In this ascent the wind on the earth was S.E. At the height of 1,300 feet the balloon entered a strong S.W. current. This direction continued up to 4,000 feet, when the wind was from the S. At the height of 8,000 feet the wind changed to S.S.W., and afterwards to S.S.E. At 11,000 feet we met with fine granular snow, and passed through snow on descending, till within 8,000 feet of the earth. We entered clouds at 7,000 feet, and passed out of them at 6,000 feet into mist.

A warm current of air was met with, of more than 3,000 feet in thickness, moving from the S.W.; that is to say, in the direction of the Gulf Stream. This was the first time a stream of air of higher temperature than on the earth had been encountered. Above this the air was dry, and higher still very dry. Fine granular snow was falling into this current of warm air.

The meeting with this S.W. current is of the highest importance, for it goes far to explain why England possesses a winter temperature so much higher than is due to our northern latitudes. Our high winter temperature has hitherto been mostly referred to the influence of the Gulf Stream. Without doubting the influence of this natural agent, it is necessary to add the effect of a parallel atmospheric current to the oceanic current coming from the same regions—a true aerial Gulf Stream. This great energetic current meets with no obstruction in coming to us or to Norway, but passes over the level Atlantic without interruption from mountains.

It cannot, however, reach France without crossing Spain and the lofty range of the Pyrenees, and the effect of these cold mountains in reducing its temperature is so great, that the former country derives but little warmth from it.

I will now say a few words in relation to the very exceptional temperatures met with in my ascent of April 6, 1864. This ascent had been arranged to take place as near to March 21 as possible, but the weather was so exceptionable that, although frequent attempts were made, it was not till April 6 that an ascent could be made.

On that day the balloon left Woolwich at 4:07 p.m., with a S.E. wind. In nine minutes, when at the height of 3,000 feet, we crossed over the river Thames, ascending very evenly at the rate of 1,000 feet in three minutes, till 11,000 feet was reached, at 4:37 p.m. We then descended at about the same rate, till within 1,500 feet of the earth, when we checked the rapidity of the descent, and reached the ground on the outskirts of a pine plantation in Wilderness Park, near Sevenoaks, in Kent.

This ascent is remarkable for the small decrease in temperature with increase of elevation. The temperature of the air was 45.5° on leaving the earth; *it did not decline at all* till after 300 feet had been passed, and then it decreased pretty gradually to 33° when 4,300 feet was reached; a warm current was then entered, and the temperature increased to 40° at 7,500 feet, the same as at 1,500 feet; it decreased to 34° at 8,800 feet, and then increased slowly to 37° nearly, at 11,000 feet, a temperature which had been experienced at the heights of 8,500, 6,500, and at 3,000 feet in ascending.

On descending, the temperature increased about 9° in the first 1,000 feet; and after remaining at about this temperature till within 7,000 feet of the earth, it gradually decreased to 40° at 3,000 feet, remained at about this point till within 1,500 feet of the earth, and then increased to 46° on the ground.

Our course in this ascent was most remarkable. After passing over the Thames into Essex, we must have re-crossed the river, and moved in an entirely opposite direction till we approached the earth again, when our direction was the same as at first.

The temperatures met with on June 13, 1864, are also very remarkable. On this occasion the balloon left the grounds of the Crystal Palace at seven o'clock. The sky was cloudless and the air perfectly clear, excepting in the direction of London. An elevation of 1,000 feet was reached in one and a quarter minutes and 3,000 feet at 7:08 p.m., when the balloon began to descend, and passed down to 2,300 feet by 7:13 p.m.; on re-ascending 3,400 feet was gained at 7:20 p.m.; after taking a slight dip the balloon again ascended to 3,550 feet (the highest point) by 7:28 p.m.; it then descended to 2,500 feet, and, after several small ascents, began the downward journey at 7:50 p.m. from the height of 2,800 feet, and reached the ground at East Horndon, five miles from Brentwood, at 8:14 p.m.

The temperature of the air on the ground before starting was 62°, declining evenly with increase of elevation till 3,000 feet was reached, when it was 51.5°; on descending, the temperature was found to be 54° at 2,300 feet; the balloon then re-ascended, the temperature declining gradually to 3,100 feet, when it began to increase, reaching 49° at 3,450 feet, above which height it declined to 47° at 3,540 feet; on again descending it increased evenly, till at 2,700 feet the thermometer read 51°, and remained about the same for 200 feet; on re-ascending the temperature scarcely differed from 51° till 3,000 feet was gained, when a sudden decrease of 2° occurred in the following 35 feet; then began our final descent, the temperature remaining the same for 400 feet; it then increased to 51.5° at 2,000 feet, and to 53.2° at 1,800 feet, below which there was scarcely any alteration till the earth was reached. This fact of no change in the temperature of the air at the time of sunset was very remarkable, for it indicated that if such on this occasion was not an accidental circumstance, the law of decrease of temperature with increase of elevation might be reversed at night for some distance from the earth.

From all the experiments it appeared that the change of temperature near the earth varied greatly in different ascents, and followed no constant law. It no doubt depended on the time of day; but the ascents were so few in number and so irregularly scattered over the months of the year, that I was unable to determine the law even approximately.

The great Captive Balloon at Ashburnham Park seemed admirably adapted to settle this point, and M. Giffard, its proprietor, most kindly placed it at my disposal for any series of experiments I was desirous of making. The balloon on a calm day could ascend to the height of 2,000 feet, its rate of ascent and descent could be regulated at will, and it could be kept stationary at any elevation. The observations made in nearly thirty ascents are published in the Transactions of the Sections in the Report of the Meeting of the British Association at Exeter, 1869.

The numbers verify the indications of the several free ascents, viz. that the decrease of temperature with increase of elevation has a diurnal range, and depends upon the hour of the day, the changes being the greatest at midday and the early part of the afternoon, and decreasing to about sunset, when with a clear sky there is little or no change of temperature for several hundred feet from the earth, whilst with a cloudy sky the change decreases from the mid-day hours at a less rapid rate to about sunset, when the decrease is nearly uniform and at the rate of 1° in 200 feet. I was not able to take any observations after sunset; but such observations are greatly needed, as there seems to be a very great probability that the temperature at the height of 1,000 feet may not undergo a greater range of temperature during the night than during the day hours; and if this be the case, the temperature at night must increase from the ground with elevation. This inference seems to be confirmed by the after sunset observations of October 2, 1865, but it is very desirable and important that the fact should be verified or contradicted by direct experiments. The law with a clear sky may be thus represented. Take the heights as ordinates of a curve of which the corresponding changes of temperature are the corresponding abscissae (considered positive when the temperature decreases, *i.e.* so that a decrease of 10° at 1,000 feet would correspond to a point on the curve whose positive abscissa is 10 and ordinate 1,000): then the curve thus formed will be somewhat hyperbolic (for the changes are greatest near the earth), the concavity being turned towards the origin, which we may call the axis. The concavity will be greatest when the curve represents the decline of temperature at a time soon after midday; but as the afternoon advances the curve gradually closes up to and coincides with the axis at or about sunset, becoming then rectilinear; after passing this critical position, in which the temperature is uniform and equal to that on the earth for the first 1,000 feet, the curve probably becomes hyperbolic again, its concavity still being turned towards the axis, so that an increase of temperature corresponds to an increase of height, and the extreme position is reached probably at or soon after midnight, when the curve returns as before, the motion being probably nearly symmetrical on both sides of the axis, and the time of a complete oscillation twenty-four hours. These changes, however, are confined to the lower regions of the atmosphere: at heights exceeding a certain elevation varying with the season of the year, there can be no doubt that the general law shows a continuous decline with elevation.

The aneroid barometer

The first aneroid barometer which I had made for these observations read correctly at 30 inches, and 0.1 inch too high at 25 inches; the error increased to 0.7 inch at 14 inches, but decreased to 0.5 inch at 11 inches. A second aneroid read very nearly the same as the mercurial barometer, from 30 inches to 12 inches. A third graduated down to 5 inches, and, most carefully made and tested under the air pump before use, read the same as the mercurial barometer throughout the high ascent to seven miles on September 5, 1862. I have taken this instrument up with me in every subsequent high ascent, and it has always read the same as the mercurial barometer. These experiments prove that an aneroid can be made to read correctly at low pressures. I may mention that on several occasions aneroid barometers have been taken whose graduations have been too limited for the heights reached: these have not broken or become deranged by being subjected to a much less pressure than they were prepared for, but have resumed their readings on the pressure again coming within their graduations.

Blackened bulb thermometer

A dull, blackened bulb thermometer, *in vacuo*, with its bulb projecting beyond the car in such a position as to receive all the sun's rays, and to show their maximum effect, never read but a few degrees higher than the temperature of the air, whilst another placed with its bulb near to the centre of the table carrying the several instruments read several degrees higher still;

but in no instance has the blackened bulb thermometer, after leaving the earth, read as high as it did when exposed to the full rays of the sun on the earth.

The fact that the readings of a blackened bulb thermometer with its bulb projecting into space, free from the influence of anybody near to it, were lower than those of a similar thermometer placed with its bulb near to a body upon which the sun's beams are arrested, is in agreement with all similar experiments I have made. In my paper "On the Radiation of Heat from the Earth," published in the *Philosophical Transactions* for 1847, I have remarked that a thermometer with its bulb in free space, so as to be fully in the passing sunbeams at the height of 14 feet from the soil, never read higher than an instrument placed in air, shaded from the sun in the hottest day in summer.

From all these experiments it seems that the heat rays, in their passage from the sun, pass the small bulb of a thermometer, communicating very little or no heat to it. Similar results were obtained by the use of Herschel's actinometer on every occasion that I had an opportunity of using it.

The lines in the spectrum

At every examination, when the spectroscope was directed to the sun, a magnificent spectrum was seen, with very numerous lines, extending from A to far beyond H, the latter line appearing not nebulous, but made up of many very fine lines; at times no spectrum was seen, when the spectroscope was directed to the sky far from the sun.

Time of vibration of a magnet

In every ascent I made many attempts to obtain the time of vibration of a horizontal magnet at different elevations in the higher regions of the atmosphere, but I failed at every trial in the ascents on August 31, September 29, October 9, January 12, April 6, June 13, and June 20. In the ascent on June 27 there were frequent periods of ten to fifteen minutes when the car was very steady, so that I was enabled to take the time of vibration as accurately as on the ground. The results of ten different sets of observations proved undoubtedly that the time of vibration was longer than on the earth. In the ascent on August 29 the balloon was constantly revolving both in ascending and in descending, but was free from oscillation for fully a quarter of an hour at the highest point, viz. nearly three miles, and the time of vibration was again found to be longer than on the earth.

Different directions of the wind at different elevations

The balloon in almost every ascent was under the influence of currents of air in different directions. The thicknesses of these currents were found to vary greatly. The direction of the wind on the earth was sometimes that of the whole mass of air up to 20,000 feet nearly, whilst at other times the direction changed within 500 feet of the earth. Sometimes directly opposite currents were met with at different heights in the same ascent, and three or four streams of air more than once were encountered moving in different directions.

The velocity of the wind

Notwithstanding the different currents of air which caused the balloon to change its direction, and at times to move in entirely opposite directions; yet, neglecting all these and all upward and downward motion, and simply taking into account the places of ascent and descent, the distances thus measured were always very much greater than the horizontal movement of the air as measured by anemometers. It may be interesting to note some instances of this, which are as follows:—Velocity of the Wind by the Balloon, and by Robinson's Anemometer at the Royal Observatory, Greenwich.

On March 31, 1863, the balloon left the Crystal Palace, Sydenham, at 4:16 p.m., and fell at Barking, in Essex, a point fifteen miles from the place of ascent, at 6:30 p.m. Neglecting all motion of the balloon, excepting the distance between the places of ascent and descent, its hourly velocity was seven miles; the horizontal movement of the air of Greenwich, as shown by Robinson's anemometer, was five miles per hour.

On April 18 the balloon left the Crystal Palace at 1:16 p.m., and descended at Newhaven at 2:46 p.m. The distance is about forty-five miles, passed over in an hour and a half, or at the rate of thirty miles per hour. Robinson's anemometer had registered less than two miles per hour.

On June 26 the balloon left Wolverton at 1:02 p.m., and fell at Littleport at 2:28 p.m. The distance between these two places is sixty miles; the hourly velocity was therefore forty-two miles per hour. The anemometer at Greenwich registered ten miles per hour.

On July 11 the balloon left the Crystal Palace at 4:53 p.m., and fell at Goodwood at 8:50 p.m., having travelled seventy miles, or at the rate of eighteen miles per hour. The anemometer at Greenwich registered less than two miles per hour.

On July 21 the balloon left the Crystal Palace at 4:52 p.m., and fell near Waltham Abbey, having travelled about twenty-five miles in fifty-three minutes, or at the rate of twenty-nine miles per hour. The horizontal movement of the air by Robinson's anemometer was at the rate of ten miles per hour.

On September 29, 1864, the balloon left Wolverhampton at 7:43 p.m., and fell at Sleaford, a point ninety-five miles from the place of ascent, at 10:30 a.m. During this time the horizontal, movement of the air was thirty-three miles, as registered at Wrottesley Observatory.

On October 9 the balloon left the Crystal Palace at 4:29 p.m., and descended at Pirton Grange, a point thirty-five miles from the place of ascent, at 6:30 p.m. Robinson's anemometer during this time registered eight miles at the Royal Observatory, Greenwich, as the horizontal movement of the air.

On January 12, 1865, the balloon left the Royal Arsenal, Woolwich, at 2:08 p.m., and descended at Lakenheath, a point seventy miles from the place of ascent, at 4:19 p.m. At the Royal Observatory, by Robinson's anemometer, during this time the motion of the air was six miles only.

On April 6 the balloon left the Royal Arsenal, Woolwich, at 4:08 p.m. Its correct path is not known, as it entered several different currents of air, the earth being invisible owing to the mist; it descended at Sevenoaks, in Kent, at 5:17 p.m., a point fifteen miles from the place of ascent. Five miles was registered during this time by Robinson's anemometer at the Royal Observatory, Greenwich.

On June 13 the balloon left the Crystal Palace at 7:00 p.m., and descended at East Horndon, a point twenty miles from the place of ascent, at 8:15 p.m. Robinson's anemometer during this time registered seventeen miles at the Royal Observatory, Greenwich.

On August 29 the balloon left the Crystal Palace at 4:06 p.m., and descended at Weybridge at 5:30 p.m., a point thirteen miles from the place of ascent. During this time fifteen miles was registered by Robinson's anemometer at the Royal Observatory, Greenwich.

Physiological observations

The number of pulsations usually increased with elevation, as also the number of inspirations: the number of my pulsations was generally 76 per minute before starting, about 90 at 10,000 feet, 100 at 20,000 feet, and 110 at higher elevations; but the increase of height was not the only element, for the number of pulsations depended also on the health of the individual. They also, of course, varied in different persons, depending much on their temperament. This was the case, too, in respect to colour; at 10,000 feet the face of some would be of a glowing purple, whilst others would scarcely be affected. At 17,000 feet my lips were blue; at 19,000 feet both my hands and lips were dark blue; at four miles high the pulsations of my heart were audible, and my breathing was very much affected; at 29,000 feet I became insensible. From all the observations it would seem that the effect of high elevation affected everyone, but was different upon the same individual at different times.

On the propagation of sounds

It was at all times found that sounds from the earth were more or less audible, according to the amount of moisture in the air. When in clouds at four miles high, I heard a railway train; but when clouds were far below, no sound ever reached the ear at this elevation. At the height of 10,000 feet the discharge of a gun has been heard; and I believe that a sound like thunder, which we heard when at a height of 20,000 feet above Birmingham, was due to the firing of some guns that were being proved there. The barking of a little dog has been heard at the height of two miles, whilst a multitude of people shouting has not been heard at 4,000 feet. So that some notes and sounds pass more readily through the air than others.

Such are some of the results derived from observations in balloons in England; but this country is of too small an area for such experiments. Wolverhampton was chosen for its central position, but in an hour or two we were always compelled to descend; whatever part of England we start from, in one hour we may be over the sea; and if we have been this time above the clouds, ignorant whether our motion has been small or at the rate of seventy or eighty miles an hour, we must penetrate them, and then our power is gone—we have had to part with our ballast to ascend, and now we have parted with gas also; we cannot ascend again; and thus, whether our fears are groundless or no, the series of observations is limited.

Far better could the experiments be made in France, or on a large continent; and I earnestly trust that the country to which we owe the balloon will utilize yet more her great invention for increase of knowledge; and it will be indeed strange if that generous and intelligent nation, which has placed so admirable an instrument at the disposal of the learned in all countries, for exploration of the higher regions, should be behindhand in its use.

Chapter IX

The high regions

Above the clouds the balloon occupies the centre of a vast hollow sphere, of which the lower portion is generally cut off by a horizontal plane. This section is in appearance a vast continent, often without intervals or breaks, and separating us completely from the earth. No isolated clouds hover above this plane. We seem to be citizens of the sky, separated from the earth by a barrier which seems impassable. We are free from all apprehension such as may exist when nothing separates us from the earth. We can suppose the laws of gravitation are for a time suspended, and in the upper world, to which we seem now to belong, the silence and quiet are so intense that peace and calm seem to reign alone.

Above our heads rises a noble roof—a vast dome of the deepest blue. In the east may perhaps be seen the tints of a rainbow on the point of vanishing; in the west the sun silvering the edges of broken clouds. Below these light vapours may rise a chain of mountains, the Alps of the sky, rearing themselves one above the other, mountain above mountain, till the highest peaks are coloured by the setting sun. Some of these compact masses look as if ravaged by avalanches, or rent by the irresistible movements of glaciers. Some clouds seem built up of quartz, or even diamonds; some, like immense cones, boldly rise upwards; others resemble pyramids whose sides are in rough outline. These scenes are so varied and so beautiful that we feel that we could remain forever to wander above these boundless planes. But the sun, which still silvers the highest of these celestial mountains, begins already to decline.

We must quit these regions to approach the earth; our revolt against gravity has lasted long enough, we must now obey its laws again. As we descend, the summits of the silvery mountains approach us fast, and appear to ascend toward us: we are already entering deep valleys which seem as if about to swallow us up; but mountains, valleys, and glaciers all flee upward. We enter the clouds and soon see the earth; we must make the descent, and in a few minutes the balloon lies helpless and half empty on the ground.

I have said that the sky, as viewed from above the clouds, is of a deep blue colour, which deepens in intensity with increase of elevation regularly from the earth, if the sky be free from clouds, or with the increase of elevation above the clouds if they be present.

The sky, if seen through clouds, is of the same pale colour as seen from the earth, at whatever elevation the clouds may be; at the height of four miles, for instance, when we were still in cloud, the blueness of the sky was of the same pale colour as it is when seen from the earth.

When the sky is free from cloud, and but little water is present in the invisible shape of vapour, the colour deepens to an intense Prussian blue at the highest elevation.

Speaking of the blueness of the sky, Sir David Brewster, in his paper "On the Polarization of the Atmosphere," observes: "We may conclude that 90° is, in the normal state of the atmosphere, the distance from the sun of the place of maximum polarization, and 45° the corresponding angle of incidence."

This determination of the place and angle of maximum polarization affords a highly probable explanation of the azure colour of the sky. Sir Isaac Newton regarded this colour as a blue of the first order, though very faint. Professor Clausius considers the vapours to be vesicles or bladders, and ascribes the blue colour of the first order to reflection from the thin pellicle of water.

In reference to these opinions the following facts are important:— 1. The azure colour of the sky, though resembling the blue of the first order when the sky is viewed from the earth's surface, becomes an exceedingly deep Prussian blue as we ascend, and, when viewed from the height of six or seven miles, is a deep blue of the second or third order. 2. The maximum polarizing angle of the atmosphere, 45°, is the same as that of air, and not that of water, which is 53.3°. At the greatest height to which I have ascended, namely, at the height of five, six, and seven miles, where the blue is the brightest, the air is almost deprived of moisture. Hence it follows that the exceedingly deep Prussian blue cannot be produced by vesicles of water, but must be caused by reflection from the air, whose polarizing angle is 45°. The faint blue which the sky exhibits at the earth's surface is, therefore, not the blue of the first order, but merely the blue of the second or third order rendered paler by the light reflected from the aqueous vapour in the lower regions of the atmosphere.

To appreciate all the beauty of cloud scenery when the air is loaded with moisture, an aerial voyage must be made on an autumn morning before sunrise, when the atmosphere is charged with the vapours of night.

The accidental circumstance of a late descent at night determined us to anchor the balloon and re-ascend before sunrise the following morning, and thus enable us to view the clouds under these conditions. It was towards the end of August, and we left the earth at 4:30 a.m. The morning was dull, warm, and misty, and the sky was covered with cloud. The balloon bore us gently upward, making the first 1,000 feet in eight minutes; all below was thick mist, veiling the surface of the earth. At 3,500 feet, still gently rising, we entered a bed of cumulo-stratus. Fifteen minutes after we left the earth we had reached 5,000 feet, and then just emerged above the clouds. They, however, presently again formed all round and above the car, closing everything from view excepting only a line, bright as silver, which indicated the east. We were in a basin of cloud whose sides extended far above us all round. We slowly rose, and when we reached its boundary the sun rose, flooding with light the whole extent of cloudland beyond, which glistened like a golden lake under his beams. The scene all round possessed a reality and grandeur, far exceeding sunrise as viewed from the earth. Grouped around the car, both above and below, there were clouds of Alpine character, sloping to their bases in glistening light, or towering upwards in sheets of shining vapour, which added the charm of contrast to the splendid tints of sunrise. The clouds spread around us like an ocean, and, continually changing their forms, suddenly gathered themselves into mountain heaps and closed all round us, hiding the sun in neutral tinted gloom; the earth was visible through breaks, and the early morning mists were seen creeping upon its surface as the daylight gathered strength.

We threw out ballast, and, after rising through the noble valleys which formed and vanished so rapidly and in so fairy-like a manner, saw the sun as it were rise again, this time flooding the atmosphere with a brilliant sea of light; and as we rose higher and left the clouds far below, we looked down upon them bathed in a golden glow of the richest hue.

Appearance of the earth viewed from a balloon

All perception of comparative altitudes of objects on or near the ground is lost—houses, trees, the undulation of the country, etc., all are reduced to one level, and even the lower detached clouds appear to rest on the earth; everything, in fact, seems to be on the same level, and the whole has the appearance of a plane. Everything seen, looking downwards from a balloon, including the clouds, seems projected upon the one visible plane beneath.

Always, however great the height of the balloon, when I have seen the horizon it has roughly appeared to be on the level of the car—though of course the dip of the horizon is a very appreciable quantity—or the same height as the eye. From this one might infer that, could the earth be seen without a cloud or anything to obscure it, as that point of the plane beneath is directly under the eye, and the boundary line of the plane approximately the same height as the eye, the general appearance would be that of a slight concavity; but I have never seen any part of the surface of the earth other than as a plane. Towns and cities, when viewed from the balloon, are like models in motion. I shall always remember the ascent of October 9, 1863, when we passed over London about sunset. At the time when we were 7,000 feet high, and directly over London Bridge, the scene around was one that cannot probably be equalled in the world. We were still so low as not to have lost sight of the details of the spectacle which presented itself to our eyes; and with one glance the homes of 3,000,000 people could be seen, and so distinct was the view, that every large building was easily distinguishable. In fact, the whole of London was visible, and some parts most clearly. All round, the suburbs were also very distinct, with their lines of detached villas, imbedded as it were in a mass of shrubs; beyond, the country was like a garden, its fields, well marked, becoming smaller and smaller as the eye wandered farther and farther away. Again looking down, there was the Thames, throughout its whole length without the slightest mist, dotted over in its winding course with innumerable ships and steamboats, like moving toys. Gravesend was visible, also the mouth of the Thames and the coast around as far as Norfolk. The southern shore of the mouth of the Thames was not so clear, but the sea beyond was seen for many miles; when at a higher elevation, I looked for the coast of France, but was unable to see it. On looking round, the eye was arrested by the garden-like appearance of the county of Kent, till again London claimed yet more careful attention.

Smoke, thin and blue, was curling from it, and slowly moving away in beautiful curves, from all except one part, south of the Thames, where it was less blue and seemed more dense, till the cause became evident; it was mixed with mist rising from the ground, the southern limit of which was bounded by an even line, doubtless indicating the meeting of the subsoils of gravel and clay. The whole scene was surmounted by a canopy of blue, everywhere clear and free from cloud, except near the horizon, where a band of cumulus and stratus extended all round, forming a fitting boundary to such a glorious view.

As seen from the earth, the sunset this evening was described as fine, the air being clear and shadows sharply defined; but, as we rose to view it and its effects, the golden hues increased in intensity; their richness decreased as the distance from the sun increased, both right and left; but still as far as 90° from the sun, rose-coloured clouds extended. The remainder of the circle was completed, for the most part, by pure white cumulus of well-rounded and symmetrical forms.

I have seen London by night. I have crossed it during the day at the height of four miles. I have often admired the splendour of sky scenery, but never have I seen anything which surpassed this spectacle. The roar of the town heard at this elevation was a deep, rich, continuous sound—the voice of labour. At four miles above London, all was hushed; no sound reached our ears.

In conclusion, let us take the balloon as we find it, and apply it to the uses of philosophy; let us make it subservient to the purposes of war, an instrument of legitimate strategy; or employ it to ascend to the verge of our lower atmosphere, and, as it is, the balloon will claim its place among the most important of human inventions, even if it remain an isolated power, and should never become engrafted as the ruling principle of the mechanism we have yet to seek in the solution of the problem of aerial navigation.

The application of the balloon, as an instrument of vertical exploration, presents itself to us under a variety of aspects, each of which is fertile in suggestion. If we regard the atmosphere as the great laboratory of changes which contain the germ of future discoveries, to belong to the chemist, the meteorologist, and the physicist, its relation to animal life at different heights, and the form of death which at certain elevations waits to accomplish its destruction; the effect of diminished pressure upon individuals similarly placed; the comparison of experiences in mountain ascents with the experiments in balloon ascents—are some of the questions which suggest themselves, and indicate the direction of inquiries which naturally ally themselves as objects of balloon investigations; sufficiently varied and important, they will be seen, to give the balloon a place as a valuable aid to the uses of philosophy.

I should wish, before closing my own portion of this work, to express the gratification I feel that French gentlemen have united with me in collecting the results of other labours in scientific research, and I hope that my experiments may be of use in future inquiries. I most willingly place my experiences at the service of any aeronaut, and hope that the time is not distant when my experiments will be surpassed by others more extensive, and that the progress of aerial navigation may give a new scope to scientific research in the balloon.

James Glaisher (1809-1903) was an English meteorologist, aeronaut, and astronomer who worked at the Royal Observatory, Greenwich for forty-three years. He was a founding member of the Royal Meteorological Society (1850) and the Aeronautical Society of Great Britain (1866). He is most famous as a pioneering balloonist and his ascent on September 5, 1862, along with Henry Tracey Coxwell, broke the world record for altitude.

Professor Liz Bentley, Chief Executive at the Royal Meteorological Society, has been a meteorologist for over twenty-five years. She has worked at the Met Office, BBC Weather Centre and the Ministry of Defence. She joined the Royal Meteorological Society as Head of Communications in 2008 and in 2010 founded theWeather Club to promote appreciation and understanding of the weather and climate to people from all walks of life.